Synthesis Lectures on Electrical Engineering

This series of short books covers a broad spectrum of titles of interest in electrical engineering that may not specifically fit within another series. Books will focus on fundamentals, methods, and advances of interest to electrical and electronic engineers.

Amr Adly · Salwa Abd-El-Hafiz

Unconventional Performance Oriented Power Transformers Design Methodologies

Springer

Amr Adly [ID]
Electrical Power Engineering Department
Cairo University
Giza, Egypt

Salwa Abd-El-Hafiz [ID]
Engineering Mathematics and Physics
Department
Cairo University
Giza, Egypt

ISSN 1559-811X   ISSN 1559-8128 (electronic)
Synthesis Lectures on Electrical Engineering
ISBN 978-3-031-85220-6   ISBN 978-3-031-85221-3 (eBook)
https://doi.org/10.1007/978-3-031-85221-3

This Springer imprint is published by the registered company Springer Nature Switzerland AG
The registered company address is: Gewerbestrasse 11, 6330 Cham, Switzerland

If disposing of this product, please recycle the paper.

To Mohammed Adly

Preface

This book aims to offer practical and efficient methodologies to identify the main power transformer leading design variables that guarantee meeting a set of required performance specifications. The book contents build on available literature related to power transformers design methodologies, a set of the lead author's lectures to senior power engineering students, in addition to a list of previous publications covering many aspects of the proposed book.

An important feature of this book is the fact that it offers practical and unconventional transformer design methodologies of which some were tested and utilized in a power transformers manufacturing plant. The methodologies merge expertise in power engineering, electromagnetic field calculations and computational intelligence to achieve the required operating performance and specifications of power transformers. Target readers are senior power engineering students as well as design office engineers working in power transformers manufacturing plants.

Giza, Egypt

Amr Adly
Salwa Abd-El-Hafiz

Competing Interests The authors have no relevant financial or non-financial interests to disclose.

The authors have no conflicts of interest to declare that are relevant to the content of this article.

All authors certify that they have no affiliations with or involvement in any organization or entity with any financial interest or non-financial interest in the subject matter or materials discussed in this manuscript.

The authors have no financial or proprietary interests in any material discussed in this article.

Contents

Introduction

It is well known that power transformers are regarded among the most important components in power systems networks. This clearly suggests that design, manufacturing and deployment of highly optimized and efficient transformer units are crucial for the minimization of the overall network capital cost and net losses [1, 2]. Moreover, power transformer manufacturers are facing increasing competitive cost and environmental challenges due to the industry market globalization. Those challenges are sometimes augmented by very specific design requests to accommodate special design specifications [3, 4]. This further highlights the importance of achieving power transformer design methodologies that can lead to meeting the required operational specifications.

Most of the available literature related to power transformers focus on operational and performance aspects (refer, for instance, to [5, 6]). Alternatively, available literature focusing on design methodologies may be classified into two categories. The first are those adopting a mixed approach of analytical design equations and empirical formulae. While this approach does not require massive computation resources, it usually involves trial-and-error schemes to achieve a set of required specifications (refer, for instance, to [7, 8]). No guarantee, however, that meeting the specifications using this methodology would be corresponding to a minimum transformer manufacturing cost. The second power transformer design category is the one adopting detailed electromagnetic field analysis approaches, especially those employing accurate finite-element-analysis (FEA) software packages (refer, for instance, to [9–11]). There is no doubt that this design methodology is more accurate and can lead to meeting the required performance specifications at the minimum transformer manufacturing cost. Nevertheless, this approach requires massive computational resources as well as costly software tools that might not be available for a considerable number of power transformers plants, especially those located in developing nations. Moreover, the approach would still involve time-consuming trial-and-error iterative computational runs. It should be mentioned here that fast and approximate design detail assessment could be extremely useful for a manufacturer price quotation as well as to minimize the necessary time-consuming trial-and-error computations using rigorous FEA software packages.

This book presents fast specifications-oriented initial design methodologies that incorporate some guidelines inferred from relevant electromagnetic field analysis studies. More specifically, the proposed methodologies are analytical in nature and, consequently, do not require massive computational resources. Yet, these methodologies implicitly incorporate some design parameter restrictions inferred from more accurate studies involving electromagnetic field computations. An important feature of the methodologies presented in this book is the introduction of some techniques that incorporate a hybrid of non-traditional heuristic and/or evolutionary computation design strategies. Since the book is focused on the main design details, tap changers are not taken into consideration.

In the following chapters of this book, different aspects related to the main power transformer design components and methodologies are discussed. More specifically, main materials used in power transformers design are reviewed in Chap. 2. In Chap. 3, important relevant basics related to magnetic circuits and electromagnetic field analysis are reviewed. Basics of power transformers design, highlighting the important correlations between rating and dimensions, are presented in Chap. 4. Approaches and trade-offs related to achieving power transformer specifications as well as the role of electromagnetic field calculations in achieving accurate power transformer design are discussed in Chap. 5. In Chap. 6, conventional analytical and semi-analytical trial-and-error computer-aided-design (CAD) approaches that may be utilized to meet requested power transformer design specifications are presented. Unconventional CAD approaches involving artificial neural networks (ANNs) and multi-objective particle swarm optimization (MOPSO) are presented in Chap. 7. Finally, some conclusions are listed in Chap. 8. It is worth mentioning that the content and reasoning included in this book would be extremely useful to senior electrical power engineering students as well as engineers working in the design departments of power transformer manufacturing establishments.

References

1. Heathcote, M. J. (1998). *The J & P transformer book: A practical technology of the power transformer*. Newnes.
2. De Almeida, A., Santos, B., & Martins, F. (2016). Energy-efficient distribution transformers in Europe: Impact of Ecodesign regulation. *Energy Efficiency, 9*, 401–424.
3. Amoiralis, E. I., Tsili, M. A., & Kladas, A. G. (2011). Power transformer economic evaluation in decentralized electricity markets. *IEEE Transactions on Industrial Electronics, 59*(5), 2329–2341.

4. Zu, E., Lu, C., Shi, W., Jui-Chan, H., & Li, H. (2020). Competitive advantage of Japan and Taiwan transformer industry. *Revista Argentina de Clínica Psicológica, 29*(5), 116.

5. Tenbohlen, S., Coenen, S., Djamali, M., Müller, A., Samimi, M. H., & Siegel, M. (2016). Diagnostic measurements for power transformers. *Energies, 9*(5), 347.

6. Sousa, J. C., Saavedra, O. R., & Lima, S. L. (2017). Decision making in emergency operation for power transformers with regard to risks and interruptible load contracts. *IEEE Transactions on Power Delivery, 33*(4), 1556–1564.

7. Sawhney, A. K. (1999). *A course in electrical machine design, dhanpat rai & co.* Publications.

8. McLyman, C. W. T. (2004). *Transformer and inductor design handbook.* CRC press.

9. Tsili, M. A., Kladas, A. G., & Georgilakis, P. S. (2008). Computer aided analysis and design of power transformers. *Computers in Industry, 59*(4), 338–350.

10. Chitaliya, G. H., & Joshi, S. K. (2013). Finite element method for designing and analysis of the transformer-A retrospective. In *International conference on recent trends in power, control and instrumentation engineering PCIE* (pp. 54–58). Citeseer.

11. Constantin, D., Nicolae, P. M., & Nitu, C. M. (2013). 3D Finite element analysis of a three phase power transformer. In *Eurocon 2013* (pp. 1548–1552). IEEE.

Common Materials Used in Power Transformers

2

This chapter is devoted to highlighting the main active and common materials used in fabricating power transformers as well as their most important characteristics. More specifically, the chapter focusses on the main characteristics of conductors used in windings, magnetic materials used in cores, and insulators. While there are different available options for each of the aforementioned materials that vary in quality and price, the choice of the most appropriate material is usually based upon the design and market limitations. This fact will be discussed and clarified through the content of this as well as subsequent chapters.

2.1   Conductors

In general, conducting materials used for transformer windings should possess a number of properties of which the most important are:

- Good electrical properties: low resistivity and low value of resistance temperature coefficient
- Good mechanical properties: high tensile strength and high ductility
- Other important properties: high melting point, corrosion resistant and easy to weld

Obviously, there is no doubt that the most widely used winding material is copper due to its superior properties. The best second option is aluminum, which is sometimes used in low rating distribution transformers having few low-voltage winding turns that could

© The Author(s), under exclusive license to Springer Nature Switzerland AG 2025   5
A. Adly and S. Abd-El-Hafiz, *Unconventional Performance Oriented Power
Transformers Design Methodologies*, Synthesis Lectures on Electrical Engineering,
https://doi.org/10.1007/978-3-031-85221-3_2

Table 2.1 Main properties of copper and aluminum conductors

Properties	Copper	Aluminum
Resistivity at 20 °C	0.0172 Ω/m/mm^2	0.0269 Ω/m/mm^2
Density at 20 °C	8933 kg/m^3	2689.9 kg/m^3
Resistivity temperature coefficient	0.393% per °C	0.4% per °C
Tensile strength	124 MPa	46.5 MPa
Melting point	1084.88 °C	660.2 °C
Coefficient of linear expansion	16.7 μm/(m °C)	23.86 m/(m °C)
Thermal conductivity	398 W/(m °C)	210 W/(m °C)

be in the form of wide sheets rather than coils. Table 2.1 summarizes the main relevant properties of copper and aluminum (refer, for instance, to [1, 2]).

It should be mentioned here that superconductors may be used in the windings of very special transformers where a high-power rating per volume ratio is required at the expense of both the construction and operation costs [3]. This is because while safe current carrying capacities for copper and aluminum are roughly estimated to be 3 and 1 A/mm^2, this value can reach more than 100 A/mm^2 for some superconducting wires [4]. Hence, usage of superconducting wires can dramatically reduce the dimensions and volume of the transformer core and, consequently, its overall dimensions. Nevertheless, operating a power transformer equipped with superconducting windings cryogenic cooling fittings and liquid Helium (or Nitrogen) supply adds both sophistication and cost to the design, construction and operation of such transformers.

2.2 Soft Magnetic Materials

In general, magnetic materials used to construct a transformer core should ideally have infinite magnetic permeability, zero coercivity, zero conductivity, high Curie point and very high saturation flux density. In other words, aside from having infinite conductivity, the ideal flux density (B) versus magnetic field (H) curve of a magnetic material used in power transformers should be as shown in Fig. 2.1. In this figure, H_c, B_m and B_r represent the coercive field, the saturation flux density, and the residual flux density, respectively.

In reality, however, no ideal material exists. A typical flux density B–H curve of any magnetic material is shown in Fig. 2.2. In this figure, H_m represents the field at which maximum saturation flux density B_m is achieved.

Generally speaking, magnetic materials used in electrical power apparatus are classified as either soft magnetic materials or hard magnetic materials. This classification is simply related to the value of the material coercive field H_c. More specifically, magnetic materials having high H_c values are classified as hard magnetic materials while those having low

Fig. 2.1 An ideal power transformer magnetic material *B–H* curve

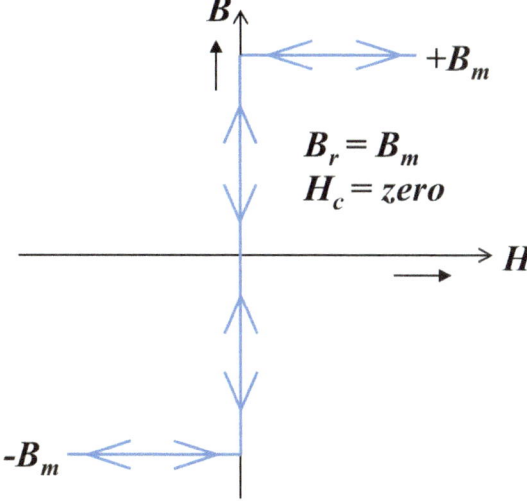

Fig. 2.2 A typical *B–H* curve of a magnetic material

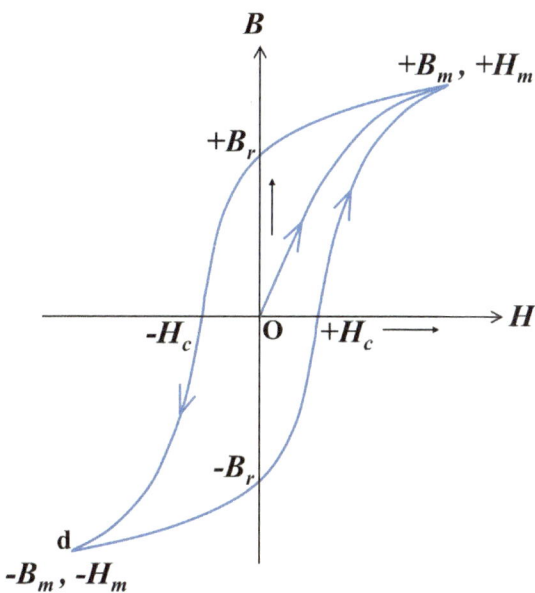

H_c values are soft magnetic materials. Referring to the ideal *B-H* curve shown in Fig. 2.1, it is obvious that power transformer cores are made of soft magnetic materials.

Since a power transformer is utilized to step up or down alternating current supplies, its core is thus subject to cyclic flux density waveforms. Hence, a hysteresis loop similar to the one depicted in Fig. 2.2 is traced (refer to [5]). In other words, every second the loop area will be traced a number of times equivalent to the supply frequency f. This is translated to an unfavorable power loss, mostly as heat energy, referred to as the hysteresis loss P_h given by [1, 6]:

$$P_h \propto f\,(Area_{B-H}),\tag{2.1}$$

$$P_h \approx K_h f B_{max}^n,\tag{2.2}$$

where $Area_{B-H}$ is the area traced by the hysteresis loop for a maximum applied cyclic flux density B_{max}, K_h is a hysteresis loss constant, and n is determined by the squareness of the traced hysteresis loop and is usually in the order of 1.5–2.

Going back to the ideal core material, the most appropriate soft magnetic material utilized in this industry is iron. Although there are many grades of iron utilized in the power transformer industry, the maximum expected flux density value is limited to 2 T. Unfortunately, to the contrary of an ideal transformer core, the conductivity σ of iron is high and is in the order of 10^7 (ohm-m)$^{-1}$. Given the fact that a power transformer core is subject to alternating magnetic flux, eddy currents induced in accordance with Faraday's law will flow in a pattern orthogonal to the applied flux as shown in Fig. 2.3. The eddy currents will result in ohmic losses inside the conducting core, mostly transformed into heat as well, referred to as the eddy current losses. For a core whose thickness d is much smaller than its length and width and where the alternating flux is parallel to its long dimensions, the eddy current losses P_e may be expressed in the form [1, 5]:

$$P_e \propto \sigma f^2 B_{max}^2 d^2,\tag{2.3}$$

$$P_e \approx K_e \sigma f^2 B_{max}^2 d^2,\tag{2.4}$$

where, once more, B_{max} is the maximum value of the applied cyclic flux density while K_e is an eddy current loss constant.

Minimization of the eddy current losses of transformer cores has been historically dealt with by addressing the thickness and conductivity parameters. For the thickness parameter, transformer cores are assembled from steel sheets which are electrically isolated from each other by an insulating surface film (such as varnish). Hence, an intended core of overall thickness d may be assembled from n thin laminations, each having a thickness equivalent to d/n as shown in Fig. 2.4. In this case, the eddy current loss of the laminated core will be a fraction of the corresponding eddy losses of the solid core as given by the following equation:

$$\frac{P_{e-laminations}}{P_{e-solidcore}} = \frac{n\left(\frac{d}{n}\right)^2}{d^2} = \frac{1}{n}.\tag{2.5}$$

Fig. 2.3 Flow patterns of eddy
currents in a conducting core,
whose thickness is much
smaller than its length and
width, as a result of orthogonal
time varying flux

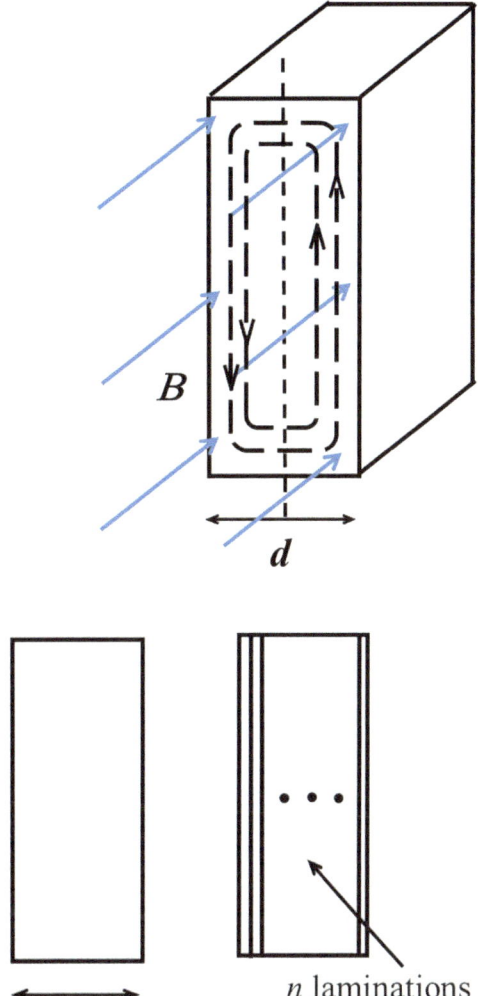

Fig. 2.4 Assembly of
transformer core of thickness
d using n isolated laminations
having a thickness d/n each

For the conductivity, on the other hand, it was found that an iron-silicon (Fe-Si) alloy
having a very small percentage (usually under 4%) of silicon content will result in a
significant reduction of conductivity without negatively affecting the magnetic properties.
In specific, the conductivity of Fe-Si laminations usually ranges between 15 and 20% of
the conductivity of pure iron.

The aforementioned discussion clearly clarifies the reason why laminated Fe-Si sheets
are used to construct power transformer cores. It is worth mentioning that the fabrication
of Fe-Si sheets is carried out using either a cold or hot rolling manufacturing process. Produced sheets are accordingly classified as cold rolled or hot rolled Fe-Si sheets depending

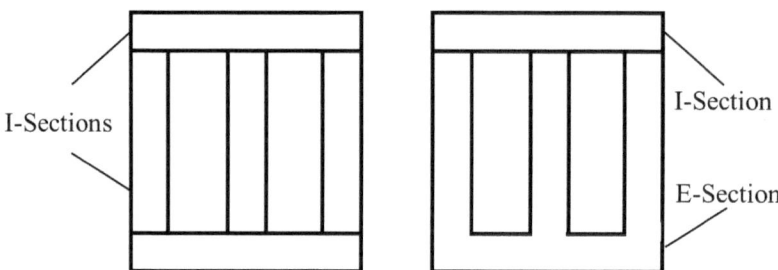

Fig. 2.5 Transformer core constructed using I- and E-sections

on their production rolling process. It is essential here to mention the most important differences between both types as those differences have impacts on their utilization and/ or application strategy (refer, for instance, to [7]).

Cold rolled steel is sometimes referred to as oriented steel. This type has the highest magnetic permeability if excited along the rolling direction. However, it is anisotropic in nature. In other words, its magnetic properties along the direction orthogonal to the rolling direction are clearly inferior to the corresponding properties along the rolling direction. Thus, this type is the favorable choice for large power transformers provided that the core is assembled from I-Sections cuttings parallel to the rolling direction (see Fig. 2.5).

Hot rolled steel, on the other hand, is sometimes referred to as non-oriented steel. It is isotropic in nature having, more or less, similar good magnetic properties along the parallel and orthogonal directions with respect to the rolling direction. However, these good magnetic properties are not as good as those of the cold rolled Fe-Si sheets along the rolling direction. For this type, the transformer core may be assembled using cuttings along the parallel and orthogonal directions with respect to the rolling direction. In other words, a core may be assembled using E- and I-Section cuttings which simplifies the core construction process (see Fig. 2.5). Nevertheless, using E-Sections usually results in increasing the percentage of wasted steel parts and is only adopted for small transformers ratings. For typical commercial types and properties of Fe-Si steel sheets, please refer to [8].

2.3 Insulating Materials

Ideal insulating materials used in the power transformer industry should possess a number of properties. More specifically, those materials should have high dielectric strength, high resistivity and high thermal conductivity. In addition, they should withstand high temperatures, resist deterioration as a result of being subjected to repeated heat cycles, and resist absorbing moisture. Moreover, they should have low dielectric loss angle and sustain bending stresses.

Table 2.2 Classification of insulating materials used in power transformers

Class	Y	A	E	B	F	H	C
Maximum temperature ($^{\circ}$C)	90	105	120	130	155	180	>180

Insulating materials used in power transformers are classified according to the maximum temperature they can withstand without degradation of their properties. Table 2.2 summarizes this classification [1]. It should be mentioned that the selection of the most appropriate insulating material is directly correlated to many factors such as the adopted design current density in the various windings as well as the cooling strategy. More specifically, proper estimation of the hottest spot temperature, which usually occurs in a center section of the low voltage winding, is essential for the proper choice of the insulating class materials (refer, for instance, to [9–11]).

References

1. Sawhney, A. K. (1999). *A course in electrical machine design, dhanpat rai & co.* Publications.
2. Olivares-Galván, J. C., De León, F., Georgilakis, P. S., & Escarela-Perez, R. (2010). Selection of copper against aluminium windings for distribution transformers. *IET Electric Power Applications, 4*(6), 474–485.
3. Berger, A., Cherevatskiy, S., Noe, M., & Leibfried, T. (2010). Comparison of the efficiency of superconducting and conventional transformers. In *Journal of physics: Conference series* (vol. 234, No. 3, p. 032004). IOP Publishing.
4. Yang, T., Li, W., Xin, Y., Chen, X., Yang, C., Tang, C., Jin, H., Hong, W., Xiong, J., Xu, J., & Li, G. (2021). Research on current carrying capacity of Bi-2223/Ag superconducting tape in the temperature range of 75–105 K. *Physica C: Superconductivity and its Applications, 582,* 1353825.
5. Adly, A. A. (2022). *Efficient unconventional models of multi-component magnetic hysteresis.* World Scientific.
6. Heathcote, M. (2011). *J & P transformer book.* Elsevier.
7. Adly, A. A., & Abd-El Hafiz, S. K. (2007). Efficient implementation of anisotropic vector Preisach-type models using coupled step functions. *IEEE Transactions on Magnetics, 43*(6), 2962–2964.
8. Tran-Cor, H. (2013). Grain oriented electrical steels. *AK Steel.*
9. Radakovic, Z., & Feser, K. (2003). A new method for the calculation of the hot-spot temperature in power transformers with ONAN cooling. *IEEE Transactions on Power Delivery, 18*(4), 1284–1292.
10. Feng, D., Wang, Z., & Jarman, P. (2014). Evaluation of power transformers' effective hot-spot factors by thermal modeling of scrapped units. *IEEE Transactions on Power Delivery, 29*(5), 2077–2085.
11. Arabul, A. Y., & Senol, I. (2018). Development of a hot-spot temperature calculation method for the loss of life estimation of an ONAN distribution transformer. *Electrical Engineering, 100,* 1651–1659.

Important Basics Related to Power Transformers

In this chapter, we review the most important basics related to power transformers theory of operation as well as some design aspects. More specifically, this chapter presents a brief review of Ampere's law, Faraday's law, Lenz's law, the magnetic circuit concept, differences between DC and AC magnetic circuits, approximation of the core B–H characteristics, the power transformer equivalent circuit, and the cooling requirements as a function of power rating.

3.1 Ampere's Law

According to Ampere's law, the circulation of magnetic field along any closed path is equivalent to the total current flowing through the surface bounded by the path as shown in Fig. 3.1 (see, for instance, [1]). Mathematically, Ampere's law may be expressed in the form:

$$\oint_{c} \overline{H} \cdot d\overline{l} = I, \tag{3.1}$$

where H is the magnetic field, c is the circular path, dl is the unit length along the circular path, and I is the current enclosed in this path.

It should be pointed out that Ampere's law is utilized in the estimation of the magnetizing current of a power transformer.

© The Author(s), under exclusive license to Springer Nature Switzerland AG 2025
A. Adly and S. Abd-El-Hafiz, *Unconventional Performance Oriented Power Transformers Design Methodologies*, Synthesis Lectures on Electrical Engineering,
https://doi.org/10.1007/978-3-031-85221-3_3

Fig. 3.1 Circulation of a magnetic field along a closed path through which current is flowing

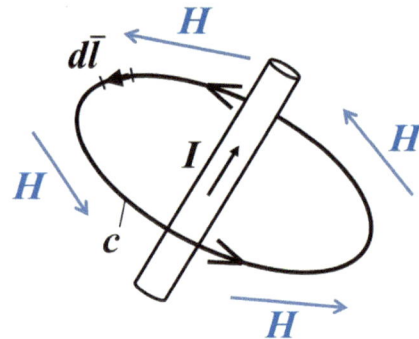

3.2   Faraday's Law and Lenz's Law

Faraday's law states that the induced voltage in a coil is proportional to the rate of variation of the magnetic flux Φ linking this coil [1, 2]. In the case when the coil circuit is closed, current will flow driven by the induced voltage. This current, according to Lenz's law, will flow such that it will create magnetic flux opposing the initial magnetic flux that induced the coil voltage [1, 2]. Faraday's law may be expressed in the following form, where the negative sign represents an account of Lenz's law:

$$ e_{ind} = -N\frac{d\Phi}{dt}, \tag{3.2} $$

where e_{ind} is the induced voltage and N is the coil number of turns.

It is worth mentioning here that (3.2) is the main formulation that correlates a power transformer winding voltage to its number of turns as well as the magnetic flux flowing in the transformer magnetic core.

3.3   The Magnetic Circuit Concept

There is no doubt that accurate estimation of the magnetic flux distribution in any power apparatus may only be achieved through rigorous electromagnetic field computation approaches. This rigorous computation requires special electromagnetic analytical expertise that may be substituted by a high-cost software tool. Both may not be available for a wide sector of technical personnel affiliated to power transformer manufacturing establishments.

For the case when approximate estimation of the magnetic flux flowing in the transformer core is acceptable, the magnetic circuit concept becomes an asset. In order to demonstrate this concept, let us consider an N-turn coil wound on a toroidal magnetic core having an average radius r and a permeability μ as shown in Fig. 3.2.

Fig. 3.2 Coil wound around a magnetic toroidal core having N turns

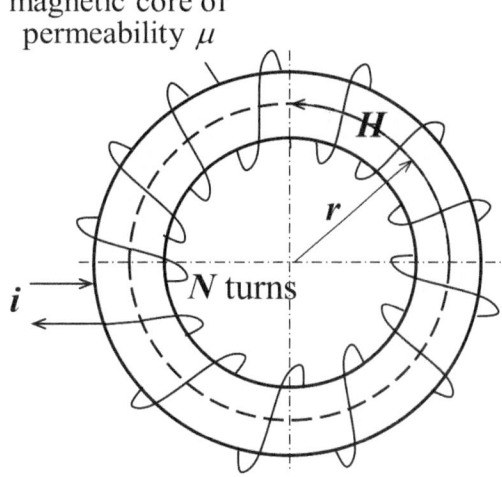

magnetic core of permeability μ

Using Ampere's law given by (3.1), the magnetic field H flowing in the toroidal core may be deduced as follows:

$$\int_0^{2\pi} Hr d\varphi = 2\pi r H = Ni \tag{3.3}$$

Hence,

$$H = \frac{Ni}{2\pi r}, \tag{3.4}$$

where i is the current flowing in the coil.

Denoting the toroidal cross-sectional area by A and expressing the field H in terms of the flux density B and the magnetic flux Φ flowing in the toroid, (3.4) may be rewritten in the form:

$$\Phi = BA = \frac{\mu A Ni}{2\pi r} = \frac{Ni}{\left(\frac{2\pi r}{\mu A}\right)}. \tag{3.5}$$

Now, consider a circular conductor as shown in Fig. 3.3 whose average radius is r connected to a voltage source E and whose conducting wire conductivity and cross-sectional area are given by σ and A, respectively. Referring to Ohm's law, the current i flowing in the conducting wire may be simply calculated from:

$$i = \frac{E}{\left(\frac{2\pi r}{\sigma A}\right)}. \tag{3.6}$$

Fig. 3.3 Circular conductor having average radius r and connected to a voltage source E

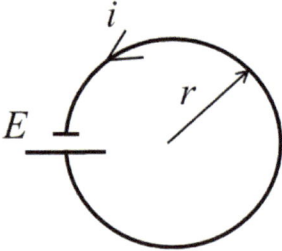

Fig. 3.4 The core and coil described in Example 3.1

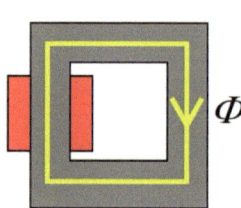

Comparing (3.5) and (3.6), it can be concluded that magnetic quantities may be (approximately) computed by analogy with electric circuits. In this concept, the magneto-motive-force Ni, the magnetic permeability μ and the magnetic flux Φ are analogous to the electro-motive-force E, the conductivity σ and the current i, respectively. In order to demonstrate this concept further, let us consider the following example. It should also be stated that the resistance of the conducting wire, given by its length divided by its conductivity and cross-sectional area, is also analogous to the so-called magnetic reluctance $\Re$ given by its length divided by its magnetic permeability and cross-sectional area.

Example 3.1 Deduce the magnetic flux density B flowing in the equal sided core of Fig. 3.4 whose cross-sectional area $A = 0.04$ m^2, permeability $\mu = 10^{-4}$, and average side length $l = 0.1$ m. The coil wound around the core has 100 turns and is carrying a current $i = 10$ A.

Solution

Using the magnetic circuit concept, the configuration shown in Fig. 3.4 may be represented by the magnetic circuit shown in Fig. 3.5.

In this circuit, the magnetic reluctance of each core side $\Re$ is given by:

$$\Re = \frac{l}{\mu A} = \frac{0.1}{(10^{-4})(0.04)} = 25000. \tag{3.7}$$

Hence, the magnetic flux may be computed using:

$$\Phi = \frac{Ni}{4\Re} = \frac{(100)(10)}{4(25000)} = 0.01 \, Wb. \tag{3.8}$$

Fig. 3.5 The magnetic circuit
analogous to the coil-core
configuration shown in Fig. 3.4

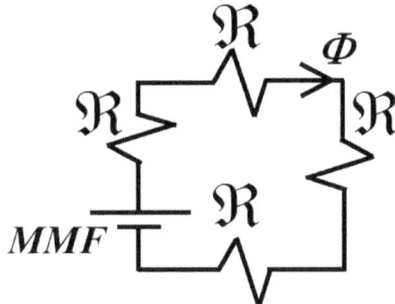

Finally, the magnetic flux density may be simply calculated from:

$$B = \frac{\Phi}{A} = \frac{0.01}{0.04} = 0.25T. \qquad (3.9)$$

3.4   Differences Between DC and AC Magnetic Circuits

While the magnetic circuit concept is handy and straight forward, it is very important
to highlight core differences between the way DC and AC magnetic circuits should be
handled. The term DC magnetic circuit refers to a magnetic circuit in which the magneto-
motive force is due to a DC electric source. Alternatively, an AC magnetic circuit refers
to the case when the magneto-motive force is due to an AC electric source.

Although the correlation between the circuit parameters of both DC and AC magnetic circuits is the same, it is important to remember that, unlike for DC magnetic
circuits, Faraday's law should be satisfied for AC magnetic circuits. More specifically,
while analyzing an AC magnetic circuit the satisfaction of Faraday's law should precede
any other analytical step. In the following two examples, the differences between DC and
AC magnetic circuits are clearly highlighted.

Example 3.2 Consider the two identical magnetic core configurations shown in Fig. 3.6,
where Core-B has a magnetic permeability μ_B equivalent to 100 times the magnetic permeability μ_A of Core-A. Both cores have equal coils and are energized by equal DC voltages
e. Compare between the currents flowing in each coil and the magnetic flux flowing in each
core.

Solution

In the absence of the need to satisfy Faraday's law and since both coils are identical and
energized by equal DC voltage sources, the currents flowing in each coil may be deduced

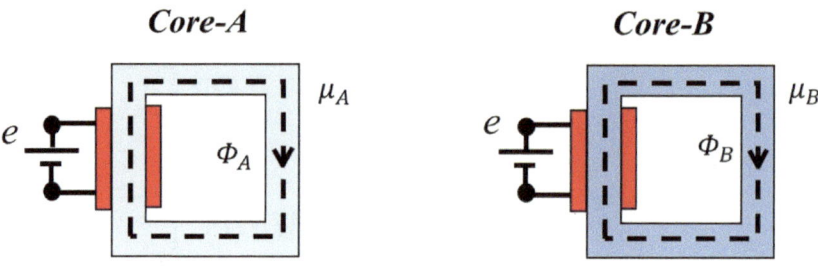

Core-A   Core-B

cross-sectional area $= a$, side length $= l$, coil resistance $= R$

Fig. 3.6 The two core configurations energized by DC source as described in Example 3.2

using Ohm's law as follows:

$$i_A = i_B = \frac{e}{R}, \tag{3.10}$$

where i_A is the current flowing in the coil of Core-A, i_B is the current flowing in the coil of Core-B, and R is the resistance of each of the identical coils.

The magnetic circuits corresponding to both configurations may, thus, be realized as shown in Fig. 3.7 where both magnetic circuits are energized by equal magneto-motive-forces.

Hence, the magnetic flux flowing in each core may be computed using:

$$\Phi_A = \left(\frac{Ne}{R}\right)\left(\frac{1}{4\Re_A}\right), \Phi_B = \left(\frac{Ne}{R}\right)\left(\frac{1}{4\Re_B}\right), \tag{3.11}$$

$$\Re_A = \frac{l}{\mu_A A}, \Re_B = \frac{l}{\mu_B A}. \tag{3.12}$$

Since $\mu_B = 100\,\mu_A$, thus:

$$\Re_A = 100\Re_B. \tag{3.13}$$

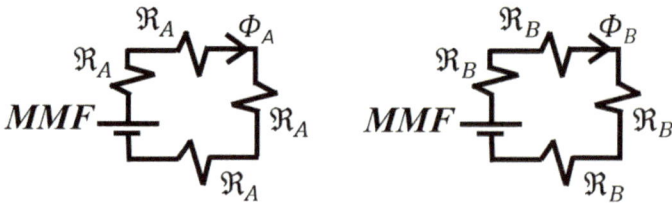

Fig. 3.7 The magnetic circuits corresponding to the core configurations depicted in Fig. 3.6

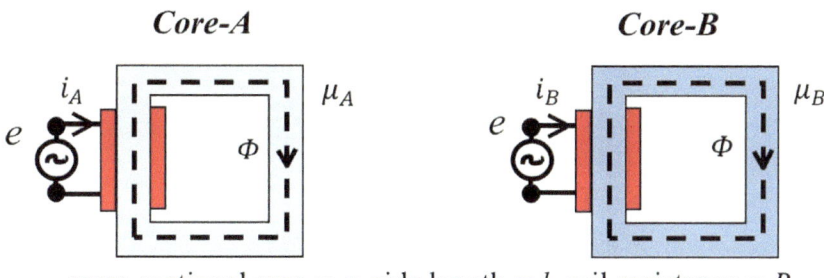

cross-sectional area = a, side length = l, coil resistance = R

Fig. 3.8 The two core configurations energized by AC source as described in Example 3.3

Consequently,

$$\Phi_B = 100\Phi_A, \tag{3.14}$$

while the supply currents are equal.

This means that for identical magnetic configurations that differ in their magnetic permeabilities while being energized through identical coils by a DC source, equal currents flow in both coils while different magnetic flux magnitudes flow in the cores. Moreover, the ratio between the magnetic flux values in both cores is equivalent to the ratio between their magnetic permeabilities.

Example 3.3 Consider the two identical magnetic core configurations shown in Fig. 3.8 where Core-B has a magnetic permeability μ_B equivalent to 100 times the magnetic permeability μ_A of Core-A. Both cores have equal coils and are energized by equal AC voltages e. Compare between the currents flowing in each coil and the magnetic flux flowing in each core.

Solution

In this example, the coils of both configurations are energized by equal AC voltage sources. Hence, Faraday's law has to be satisfied. Neglecting the coils resistances compared to their inductances and applying (3.2) to the coils of Core-A and Core-B, we get:

$$e = e_{Core-A} = e_{Core-B} = -N\frac{d\Phi}{dt}. \tag{3.15}$$

The analogous magnetic circuits for both cores will thus be as shown in Fig. 3.9.

The supply currents for both cores may, hence, be deduced from:

$$MMF_A = Ni_A = (\Phi)(4\Re_A), \; MMF_B = Ni_B = (\Phi)(4\Re_B), \tag{3.16}$$

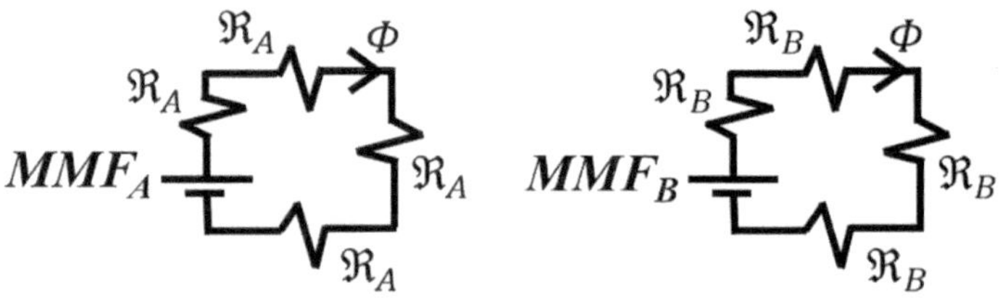

Fig. 3.9 The magnetic circuits corresponding to the core configurations depicted in Fig. 3.8

$$i_A = \left(\frac{4\Phi}{N}\right)\left(\frac{l}{a\mu_A}\right), i_B = \left(\frac{4\Phi}{N}\right)\left(\frac{l}{a\mu_B}\right). \tag{3.17}$$

Since $\mu_B = 100\ \mu_A$, thus:

$$i_A = 100 i_B, \tag{3.18}$$

while the magnetic flux flowing in both cores are equal.

Once more, this is a totally different result in comparison to the case when both cores are energized by a DC voltage source. Another very important conclusion, which is directly related to power transformers design, can be inferred. That is, employing high quality magnetic material having higher magnetic permeability will result in decreasing the current required to support a certain flux, usually referred to as the magnetizing current.

3.5   Approximate Representation of the Core *B–H* Characteristics

It has been shown in Chap. 2 that the actual *B–H* curve of a magnetic material exhibits hysteresis as shown in Fig. 2.2. Hysteresis is best defined as a multi-branch nonlinearity for which branch to branch transition occurs at local extremum values and is dependent on the magnetic field sequence history [3]. Generally speaking, dealing with the hysteresis phenomenon is sophisticated since the *B–H* relation is nonlinear and history dependent. In other words, taking the actual *B–H* relation of a magnetic material into consideration requires utilization of hysteresis models (see, for instance, [4, 5]).

In the case when the width of the *B–H* curve is small, which is fortunately the case for soft magnetic materials utilized in power transformers cores, the curve may be approximated as a nonlinear function as shown in Fig. 3.10. Please note that the curve shown in Fig. 3.10 may be subdivided into two zones; an almost linear zone for relatively low magnetic field values and a nonlinear zone for high field values. Here, the magnetic permeability μ is clearly a function of the applied magnetic field H. Highest values of the

Fig. 3.10 Approximate representation of the *B–H* curve of a soft magnetic material

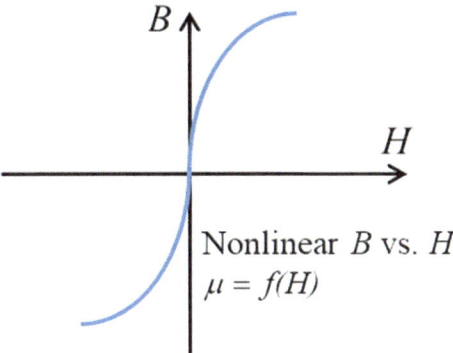

Nonlinear B vs. H

$\mu = f(H)$

Fig. 3.11 Approximate representation of the *B–H* curve of a soft magnetic material for relatively low magnetic field values

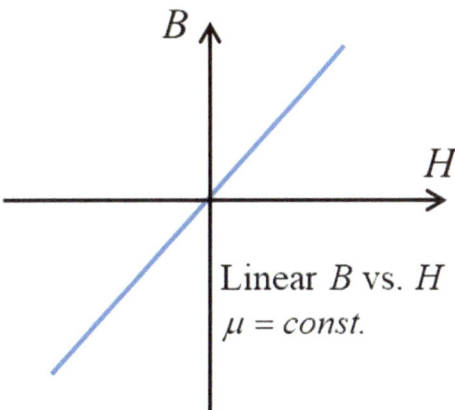

Linear B vs. H

$\mu = const.$

magnetic permeability of the material correspond to low applied field values and clearly decrease as the applied magnetic field increases. For the case when the applied magnetic field is restricted within the linear zone, the *B–H* curve may be further approximated as shown in Fig. 3.11. In this case, the magnetic permeability μ may be represented by a constant value.

3.6   Cooling Requirements for Higher Power Transformer Ratings

Comparing cooling requirements of power transformers having different ratings, it can be easily concluded that as the rating increases the cooling strategy becomes more demanding. For instance, small power transformers used for home appliances are cooled by natural convection. In other words, heat lost in small power transformers as a result of its various copper and iron losses is usually dissipated through its outer surface area by

Fig. 3.12 Two hypothetical cuboidal electrical machines having power rating ratios 1 to 1000

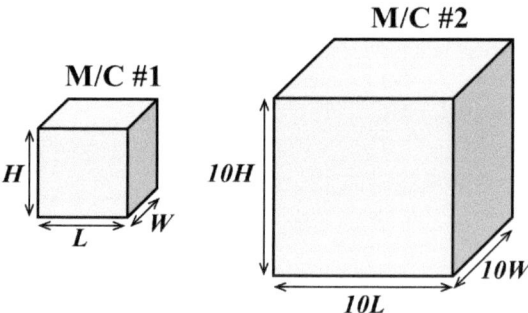

natural convection. For larger power transformers, natural heat convection is not enough. It has to be boosted by forced convection, or by placing the power transformer in oil tanks having cooling fins, or even both. As the rating gets higher more sophisticated cooling approaches, such as using Sulfur Hexafluoride (SF_6) and water cooling, have to be utilized. This fact may be justified by the following example.

Example 3.4 Consider the two hypothetical cuboidal electrical machines denoted by M/C #1 and M/C #2 shown in Fig. 3.12. Knowing that the rating of an electrical machine is, more or less, proportional to its volume and assuming similar efficiencies compare between the ratios of their losses and their surface areas. Comment on the comparison results in view of the fact that natural convection is proportional to the surface area of the machine.

Solution

Given the dimensions of both machines, it is clear that the rating of M/C #2 is equivalent to 1000 times that of M/C #1.

Assuming similar efficiencies, then, the losses in M/C #2 will also be equivalent to 1000 times those of M/C #1.

The surface area of both machines may be computed from:

$$Area_{M/C\#1} = 2(LW + WH + HL), \tag{3.19}$$

$$Area_{M/C\#2} = 200(LW + WH + HL), \tag{3.20}$$

where $Area_{M/C\#1}$ and $Area_{M/C\#2}$ are the surface areas of the first and second machines, respectively.

Hence,

$$Area_{M/C\#2} = 100 Area_{M/C\#1}. \tag{3.21}$$

In other words, although the losses M/C #2 are equivalent to 1000 times those of M/C #1, the surface area of M/C #2 is only equivalent to 100 times that of M/C #1. This clearly highlights the need for more sophisticated cooling approaches as the power rating increases. Examples of these approaches include dry type air natural (AN), dry type air forced (AF), oil natural-air natural (ONAN), oil natural-air forced (ONAF), oil forced-air forced (OFAF), in addition to more sophisticated types for extremely large uncommon ratings.

3.7   The Power Transformer Equivalent Circuit

There is no doubt that the operating performance of any power transformer is assessed through its equivalent circuit. While the equivalent circuit of a transformer is well known and may be referred to in any relevant textbook, it is important to correlate the different equivalent circuit components to their corresponding transformer physical constituents. This should shed some light on how the various materials used in a power transformer and its physical geometrical configuration could have an effect on the operating performance.

As a first step, consider the ideal single-phase power transformer shown in Fig. 3.13 at no-load. In this figure, V_p, E_p, V_s, E_s, Φ_m, and I_m represent the supply voltage, the transformer primary voltage, the load voltage, the secondary transformer voltage, the core maximum flux and the input magnetization current, respectively. Please note that, in this case, an ideal transformer is characterized by the assumptions of infinite core permeability, zero core losses, zero leakage flux from the core and zero winding resistances.

Focusing on the induced voltage rather than its polarity and referring to (3.2), it can be easily shown that:

$$e_p = N_1 \frac{d\Phi}{dt}, \tag{3.22}$$

$$e_s = N_2 \frac{d\Phi}{dt}, \tag{3.23}$$

Fig. 3.13 An ideal power transformer

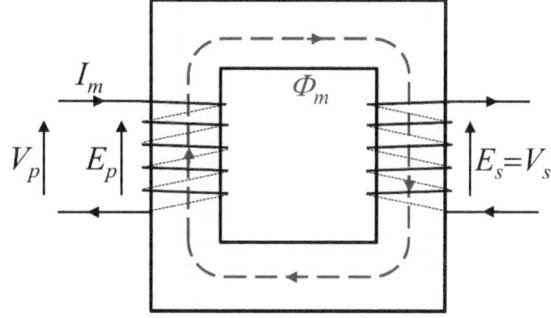

where e_p, e_s, N_1, N_2 and Φ represent the primary instantaneous voltage, the secondary instantaneous voltage, the primary coil number of turns, the secondary coil number of turns and the instantaneous core flux, respectively.

Hence,

$$\frac{e_p}{e_s} = \frac{E_p}{E_s} = \frac{N_1}{N_2}. \tag{3.24}$$

The magnetic circuit correlation between the instantaneous magnetizing current i_m and instantaneous core flux Φ, in this case, may be written in the form:

$$N_1 i_m = (\Phi)(\Re_{Core}) = (\Phi)\left(\frac{l_c}{A_c \mu_c}\right), \tag{3.25}$$

where $\Re_{core}$, l_c, A_c and μ_c represent the magnetic core reluctance, mean flux path length, cross sectional area and permeability, respectively.

Since μ_c is assumed to be infinity for the ideal transformer, no magnetizing current is required to support the flux induced in the core as a result of the applied time varying voltage.

Next, consider the ideal single-phase power transformer with the exception of assuming a realistic finite core permeability. Referring to (3.25), the magnetizing current may be computed from:

$$i_m = \left(\frac{\Phi l_c}{N_1 A_c \mu_c}\right). \tag{3.26}$$

A very important conclusion here is that for the same applied voltage, number of primary turns and core configuration, the higher the magnetic permeability the less magnetizing current is drawn.

For the case when the secondary is loaded, additional currents I_1 and I_2 flow in the primary and secondary windings as shown in Fig. 3.14 where Z is the load impedance, while Φ_p and Φ_s are the flux components linking the primary and secondary windings, respectively. Since the voltage applied to the primary winding is maintained the same, we expect from Faraday's law that the magnetic flux flowing in the core would be the same.

Fig. 3.14 A loaded ideal power transformer having finite permeability

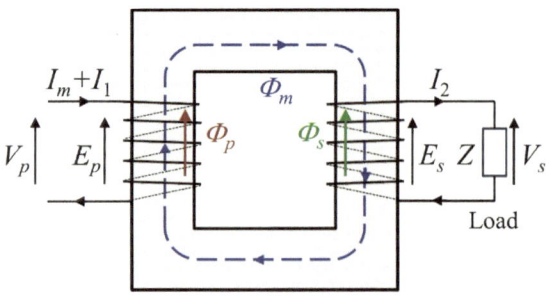

Fig. 3.15 Magnetic circuit
corresponding the power
transformer shown in Fig. 3.14

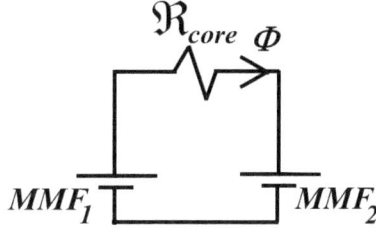

Referring to the corresponding magnetic circuit in this case as shown in Fig. 3.15, it can be concluded that:

$$MMF_1 - MMF_2 = N_1(i_1 + i_m) - N_2(i_2) = \left(\frac{\Phi l_c}{A_c \mu_c}\right), \tag{3.27}$$

where i_1 and i_2 are the instantaneous additional primary and secondary currents, respectively.

Hence, given that the flux is the same and referring to (3.27) it can be easily concluded that:

$$N_1(i_1) - N_2(i_2) = 0. \tag{3.28}$$

Thus,

$$\frac{i_1}{i_2} = \frac{I_1}{I_2} = \frac{N_2}{N_1}. \tag{3.29}$$

Referring to (3.22) for an applied sinusoidal supply voltage, it can be shown that:

$$e_p = \sqrt{2}E_p \sin(\omega t) = N_1 \frac{d\Phi}{dt}, \tag{3.30}$$

where ω is the supply angular frequency and t is the time.

Thus,

$$\Phi = \frac{\sqrt{2}E_p}{N_1} \int \sin(\omega t)dt, \tag{3.31}$$

$$\Phi = -\frac{\sqrt{2}E_p}{\omega N_1} \cos(\omega t), \tag{3.32}$$

$$\Phi = \frac{\sqrt{2}E_p}{\omega N_1} \sin\left(\omega t - \frac{\pi}{2}\right), \tag{3.33}$$

$$\Phi_m = \frac{\sqrt{2}E_p}{\omega N_1}. \tag{3.34}$$

Fig. 3.16 Equivalent circuit of an ideal transformer having finite magnetic core permeability

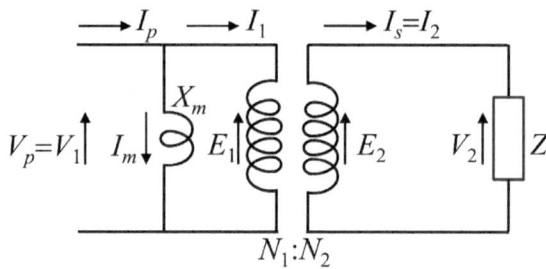

$$N_1 : N_2$$

Hence, the induced magnetic flux in a transformer core lags the applied voltage by $90°$. Since the instantaneous magnetizing current i_m is directly correlated to the instantaneous flux, the magnetizing current may be represented in a circuit component form by an inductor whose terminals are connected to the supply voltage terminals as shown in Fig. 3.16. In this figure, X_m is a fictitious reactance corresponding to the magnetizing current while E_1 and E_2 are the assumed voltages across the ideal transformer.

In the case when the core losses are further taken into consideration, the total core losses P_{core} may be deduced from (2.2) and (2.4) as follows:

$$P_{core} = P_e + P_h, \tag{3.35}$$

$$P_{core} \approx K_h f B_{max}^{(1.5-2)} + K_e \sigma f^2 B_{max}^2 d^2. \tag{3.36}$$

Adopting the upper limit in the first term of (3.36), the core losses may be formulated in the form:

$$P_{core} \approx K_{core} B_{max}^2, \tag{3.37}$$

where K_{core} is some constant dependent on the core material and configuration.

Referring to (3.34), (3.37) may be rewritten in the form:

$$P_{core} \approx K_{core} \left(\frac{\sqrt{2}}{\omega N_1 A_c} \right)^2 E_p^2. \tag{3.38}$$

It can be seen that the core losses are proportional to the square of the primary voltage irrespective of the supply current. Hence, the core losses may be represented by a fictitious resistance R_m connected across the primary voltage terminals as shown in Fig. 3.17. In this figure, I_f, I_p and I_s represent the core loss equivalent current, the overall primary current and the overall secondary current, respectively.

For actual transformers, the assumption of zero resistance is not valid. Moreover, the assumption that the flux generated by the transformer windings currents is only confined in its magnetic core is not valid as well. Actually, some flux that links each of the two windings leaks from the core and circulates outside the core as shown in Fig. 3.18. In this

Fig. 3.17 Equivalent circuit of an ideal transformer having finite magnetic core permeability and taking core losses into account

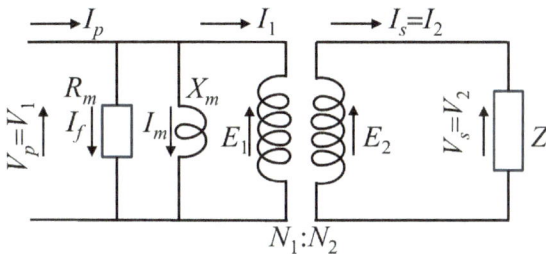

Fig. 3.18 Demonstration of the possible leakage flux in a power transformer

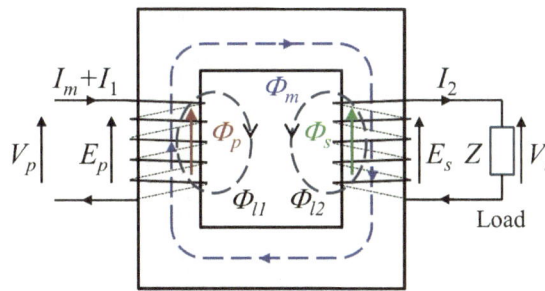

figure, the leakage flux for the primary and secondary windings is denoted by Φ_{l1} and Φ_{l2}, respectively. In such a case, the total primary and secondary windings flux Φ_p and Φ_s may be written in terms of the primary and secondary voltages as follows:

$$e_p = N_1 \frac{d\Phi_p}{dt} = N_1 \frac{d\Phi_m}{dt} + N_1 \frac{d\Phi_{l1}}{dt}, \tag{3.39}$$

$$e_s = N_2 \frac{d\Phi_s}{dt} = N_2 \frac{d\Phi_m}{dt} + N_2 \frac{d\Phi_{l2}}{dt}. \tag{3.40}$$

The last term of each of (3.39) and (3.40) may be interpreted as reactive voltage drop on each of the two transformer windings. In other words, the actual power transformer may be realized by an ideal transformer whose terminals are separated from the actual supply and load terminals by circuit components accounting for windings resistances R_1 and R_2, windings leakage reactances X_{l1} and X_{l2}, magnetizing reactance X_m and core loss resistance R_m as depicted in Fig. 3.19 (refer, for instance, to [6–8]).

Please note that the equivalent circuit shown in Fig. 3.19 corresponds to a single-phase power transformer. For three-phase power transformers, the same figure may be regarded as the equivalent circuit per phase.

Fig. 3.19 Equivalent circuit of an actual single-phase transformer

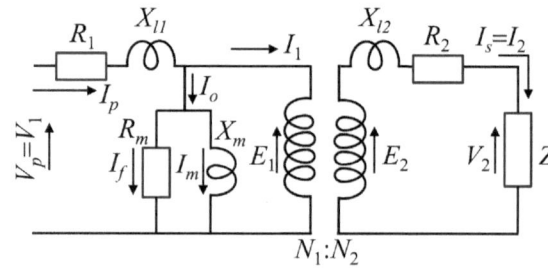

References

1. Cheng, D. K. (1989). *Field and wave electromagnetics*. Pearson Education India.
2. Haus, H. A., & Melcher, J. R. (1989). *Electromagnetic fields and energy* (Vol. 107). Prentice Hall.
3. Mayergoyz, I. (1986). Mathematical models of hysteresis. *IEEE Transactions on magnetics*, 22(5), 603–608.
4. Mayergoyz, I. D. (2003). *Mathematical models of hysteresis and their applications*. Academic press.
5. Adly, A. A. (2022). *Efficient unconventional models of multi-component magnetic hysteresis*. World Scientific.
6. Sawhney, A. K. (1999). *A course in electrical machine design, dhanpat rai & co*. Publications.
7. Heathcote, M. (2011). *J & P transformer book*. Elsevier.
8. Flanagan, W. (1993). *Handbook of transformer design and applications*. McGraw-Hill, Inc.

Basics of Power Transformers Design: Rating-Dimensions Correlations

4

When it comes to power transformer design, the main challenge is to satisfy a list of operation specifications. This challenge is translated to identifying the transformer design details that would lead to the guarantee of meeting the required specifications. Power transformer design strategies may be generally subdivided into accurate approaches that usually utilize finite element analysis (FEA) tools and analytical approaches that usually incorporate additional roughly estimated safety margin factors to guarantee the fulfilment of the required specifications even at the expense of the possibility of deviating from achieving the minimum cost design [1, 2]. Obviously, there is a trade-off between the two strategies. Accurate approaches require high computational resources, relatively expensive software tools and the full design details to assess the accurate performance. Thus, design optimization using this strategy is time consuming. On the other hand, adopting the analytical approximate design strategies does not require massive computational resources and offers relatively speedy optimization scenarios at the expense of the accuracy and/or the guarantee to achieve the minimum cost design. Traditionally, both strategies required trial-and-error iterative computational schemes. The novel contribution of this chapter, however, is to present the details of an efficient and relatively accurate non-iterative specifications-oriented design methodology.

According to this methodology, main specification requirements may be directly correlated to four main design parameters [3, 4]. More specifically, the transformer volt-ampere rating (S), the overall ohmic copper losses (P_{cu}), the no-load losses (P_{nl}), and the ohmic reactance per phase (X) may be correlated to the transformer window height (H_w), the limb diameter (D), the maximum core flux density (B_c), and the average conductors' current density (J). It should be pointed out that all other transformer design details may be

© The Author(s), under exclusive license to Springer Nature Switzerland AG 2025   29
A. Adly and S. Abd-El-Hafiz, *Unconventional Performance Oriented Power Transformers Design Methodologies*, Synthesis Lectures on Electrical Engineering, https://doi.org/10.1007/978-3-031-85221-3_4

deduced from the aforementioned design specifications and leading variables. Details of the proposed methodology are presented in the following sections of this chapter.

4.1   Single and Three Phase Transformer Core Configurations

In general, there are two types of core configurations for single and three-phase power transformers. More specifically, power transformer cores are either of the core type or shell type as shown in Figs. 4.1 and 4.2. Since core type transformers are the dominantly common ones, the proposed detailed design methodologies will be confined accordingly.

For assembly convenience, the windings of a transformer are usually wound around a holder or bobbin before being placed all together on the transformer limb (see Fig. 4.3). Depending on the voltage value, the winding process involves in addition to the insulated winding conductors, additional insulating layers as needed and possible spacers to act as cooling ducts. As a rule of thumb, the lower voltage winding is usually placed closer to the limb while the higher voltage winding occupies the outer layer of the windings set.

The core cross section should ideally be circular. Given the fact that the transformer core should be made up of laminated sheets as discussed in Chap. 2, an ideal core would be as roughly shown in Fig. 4.4. From a practical point of view, the ideally circular cross-sectional area would require a very high number of laminations having different widths. This presents a fabrication unjustified overhead and would result in a big percentage of unutilized laminated steel remnant sections.

Alternatively, the simplest core cross section would be a square one as shown in Fig. 4.5, where a is the core side length and D is the presumed core limb diameter. Correlating the actual gross core area A_{gi} to the presumed ideal circular core area, it can be shown that:

Fig. 4.1 Core and shell type single-phase transformer core configurations

Fig. 4.2 Core and shell type three-phase transformer core configurations

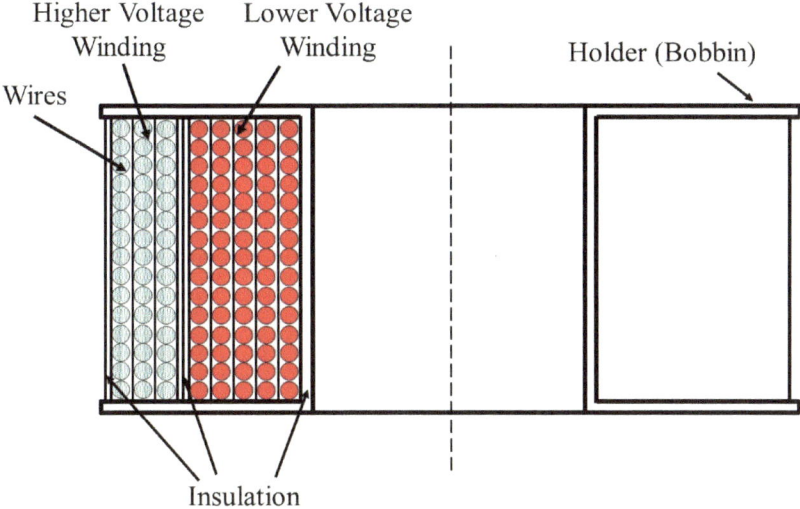

Fig. 4.3 Typical transformer windings configuration

Fig. 4.4 An ideal transformer
limb cross section

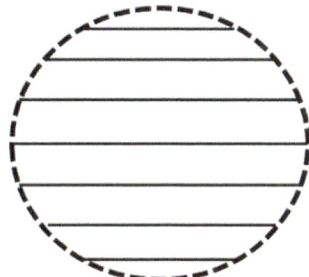

$$A_{gi} = a^2 = \left(Dcos\left(\frac{\pi}{4}\right)\right)^2 = 0.5D^2, \tag{4.1}$$

$$K_c = \frac{A_{gi}}{\frac{\pi}{4}D^2} = \frac{0.5}{\pi/4} = \frac{2}{\pi} \approx 0.6366, \tag{4.2}$$

where K_c represents the ratio between the actual gross core area and the presumed ideal circular area.

Please note that while adopting a square core cross-sectional area simplifies the construction and assembly of a power transformer, it results in an unnecessary excessive winding copper volume for the same net core area. Obviously, the higher the value of K_c, the less unnecessary copper volume is involved. In order to go one step further in maximizing the value of K_c without dramatically complicating the fabrication and assembly process, a cruciform cross-section may be adopted having long and short dimensions

Fig. 4.5 A limb core having a
square cross section

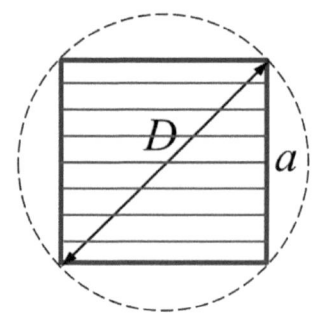

Fig. 4.6 A limb core having a
cruciform cross section

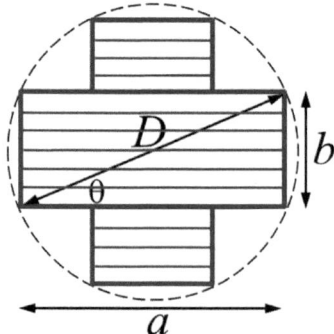

denote by a and b, respectively, as shown in Fig. 4.6. In this case, only two lamination
dimensions are required, which is quite manageable.

In this case, the gross core area A_{gi} may be given by:

$$A_{gi} = 2ab - b^2, \tag{4.3}$$

$$A_{gi} = 2D^2 \sin\theta \cos\theta - (D\sin\theta)^2, \tag{4.4}$$

where

$$\theta = tan^{-1}\left(\frac{b}{a}\right). \tag{4.5}$$

For maximum core cross-sectional area, the value θ may deduced as:

$$\frac{\partial A_{gi}}{\partial \theta} = 2D^2 \cos 2\theta - D^2 \sin 2\theta = 0, \tag{4.6}$$

$$2\cos 2\theta = \sin 2\theta, \tag{4.7}$$

$$\theta = \frac{1}{2}tan^{-1}(2) = 31.7^o. \tag{4.8}$$

Substituting (4.8) in (4.4), A_{gi} and K_c may be given by:

$$A_{gi} \approx 0.618D^2, \tag{4.9}$$

$$K_c \approx \frac{A_{gi}}{\frac{\pi}{4}D^2} \approx \frac{0.618}{\pi/4} = 0.78686. \tag{4.10}$$

Please note that this value of K_c reflects about 24% increase in the actual gross area compared to that of a square cross-section core area. Practically, transformer plants utilize a finite multi-step laminated core that might involve 20 different lamination widths or more in order to approach a K_c value of 0.9 and beyond. In such cases, more sophisticated optimization algorithms are used to identify the number and relative dimensions of each adopted lamination width (refer, for instance, to [5, 6]).

It should be pointed out that the net core cross-sectional area A_i is usually estimated to be 95% of the gross area due to the presence of laminations insulated films. Hence, the net core cross-sectional area may be given by:

$$A_i = 0.95A_{gi} = 0.95K_c\frac{\pi}{4}D^2. \tag{4.11}$$

4.2 The Voltage Equation

It has been shown in (3.34) that the applied root mean square voltage of the primary winding may be correlated to the maximum core flux, the supply angular frequency and the primary winding number of turns. By reformulating (3.34) in terms of the maximum core flux density B_c and the core net area A_i, it can be shown that:

$$\Phi_m = B_cA_i = \frac{\sqrt{2}E_p}{2\pi fN_1}, \tag{4.12}$$

$$E_p = \frac{2\pi}{\sqrt{2}}fB_cA_iN_1, \tag{4.13}$$

$$E_p = 4.44fB_cA_iN_1. \tag{4.14}$$

From (4.11), the aforementioned voltage equation may further be elaborated for primary and secondary windings of a single-phase transformer as given by:

$$V_1 = 4.44fB_cA_iN_1 = 4.44\left(0.95K_c\frac{\pi}{4}\right)fB_cD^2N_1, \tag{4.15}$$

$$V_2 = 4.44fB_cA_iN_2 = 4.44\left(0.95K_c\frac{\pi}{4}\right)fB_cD^2N_2. \tag{4.16}$$

Likewise, the voltage equations for a three-phase transformer may be given in the forms:

$$V_{ph1} = 4.44fB_cA_iN_1 = 4.44\left(0.95K_c\frac{\pi}{4}\right)fB_cD^2N_1, \tag{4.17}$$

$$V_{ph2} = 4.44fB_cA_iN_2 = 4.44\left(0.95K_c\frac{\pi}{4}\right)fB_cD^2N_2. \tag{4.18}$$

It should be pointed out that as the frequency increases, the core area and/or the number of turns may be reduced for that same applied voltage. In other words, for higher frequencies, transformers may end up having less weight, volume or both.

4.3　　The Transformer Window Space Factor

The window space factor of a transformer S_w can be simply defined as the ratio between the window area occupied by copper conductors to that of the window. This is clearly highlighted in Fig. 4.7 for single and three-phase core type transformers. In this figure, N_H is the high-voltage winding number of turns, N_L is the low-voltage winding number of turns, l_{WH} is the average windings height, H_W is the window height, W_W is the window width, while b_w and a_w represent the average winding width and spacing between the low and high-voltage windings, respectively.

Defining the high-voltage winding conductor(s) cross-sectional area by ac_H, the low-voltage winding conductor(s) cross-sectional area by ac_L, the window space factor of a

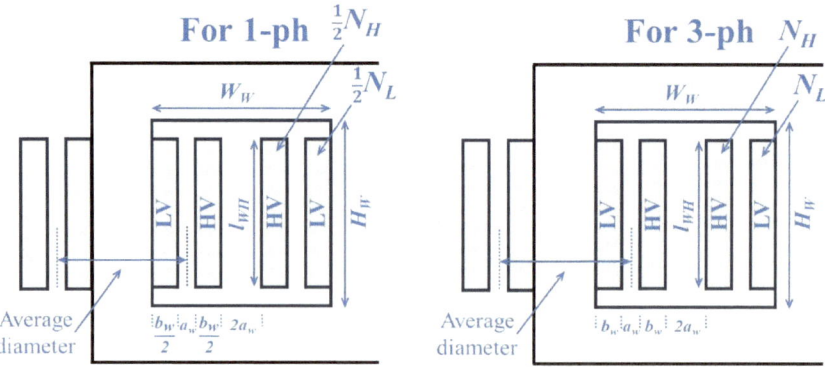

Fig. 4.7 Single and three-phase core type transformer window schematics demonstrating the windings and spacings rough configurations

single-phase transformer may be expressed in the form:

$$S_w = \frac{N_H ac_H + N_L ac_L}{H_W W_W}.$$ (4.19)

On the other hand, for a three-phase transformer, the window space factor may be given by:

$$S_w = \frac{2N_H ac_H + 2N_L ac_L}{H_W W_W}.$$ (4.20)

Denoting the high-voltage and low-voltage windings current densities by J_H and J_L, respectively, the high and low voltage conductors' cross-sectional areas of a single-phase transformer may be expressed in the form:

$$ac_H = \frac{I_H}{J_H},$$ (4.21)

$$ac_L = \frac{I_L}{J_L},$$ (4.22)

where I_H and I_L represent the high-voltage and low-voltage currents, respectively.

Similarly, the high and low voltage conductors' cross-sectional areas of a three-phase transformer may be expressed in the form:

$$ac_H = \frac{I_{phH}}{J_H},$$ (4.23)

$$ac_L = \frac{I_{phL}}{J_L},$$ (4.24)

where I_{phH} and I_{phL} represent the high-voltage and low-voltage phased currents, respectively.

Usually, $J_L \leq J_H \leq 1.3 J_L$. Assuming the ratio c_j to be the ratio between low-voltage and high-voltage windings current densities, hence:

$$J_H = c_j J_L.$$ (4.25)

It should be pointed out that the typical maximum current densities in ONAN transformers is about 2.3 Amp/mm^2, and about 3.2 Amp/mm^2 for ONAF transformers.

From (3.29), (4.23) and (4.24), we get:

$$\frac{I_{phH}}{I_{phL}} = \frac{J_H ac_H}{J_L ac_L} = \frac{N_L}{N_H}.$$ (4.26)

Substituting (4.25) into (4.26), we get:

$$\frac{I_{phH}}{I_{phL}} = \frac{c_j J_H ac_H}{J_H ac_L} = \frac{N_L}{N_H}. \tag{4.27}$$

Hence:

$$ac_L N_L = c_j ac_H N_H. \tag{4.28}$$

Substituting (4.28) into (4.19) and (4.20) for single and three-phase transformers, respectively, we get:

$$S_w = \frac{(1 + c_j) N_H ac_H}{H_W W_W}, \tag{4.29}$$

$$S_w = \frac{2(1 + c_j) N_H ac_H}{H_W W_W} \tag{4.30}$$

A rough approximation of the window space factor may be deduced from the expression [7]:

$$S_w = \begin{cases} \frac{8}{30 + HV_{inKV}} & \text{for ratings up to 10 } KVA \\ \frac{10}{30 + HV_{inKV}} & \text{for ratings up to 200 } KVA \\ \frac{12}{30 + HV_{inKV}} & \text{for ratings up to 1000 } KVA \\ 0.15 - 0.25 & \text{for higher ratings} \end{cases}, \tag{4.31}$$

where HV_{inKV} represents the transformer high-voltage rating expressed in kilo volts.

4.4   The Transformer Optimum Window Height to Width Range

It has been shown that the area of a power transformer window is mainly mandated by the design number of winding turns, cross sectional areas of the conductors and the insulation content mandated by the specified higher voltage value. Obviously, designing a power transformer involves selection of its dimensions including the window height to width ratio.

Although assessments of eddy current losses in power transformers are mainly confined to the transformer core, a smaller percentage of such losses also exists in the transformer windings as a result of time varying stray fields penetrating them (see, for instance, [8, 9]). It turns out that these losses are highly dependent on many factors including the transformer window height to width ratio. In other words, proper selection of the transformer window height to width ratio is crucial to minimize those losses and, consequently, maximize the transformer power efficiency.

In order to demonstrate this point, finite element analysis of three single-phase transformers operating at no-load and having equal window areas of 2 m^2 yet different window height to width ratios have been carried out using the Tera Analysis Quickfield software

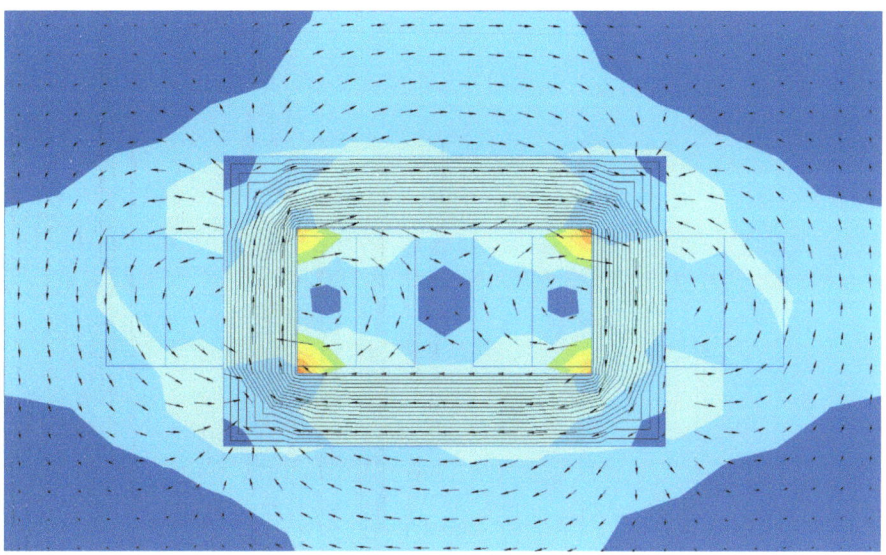

Fig. 4.8 Finite element analysis of the stray magnetic fields in a power transformer window having a window height to width ratio of 0.5

[10]. Figures 4.8, 4.9 and 4.10 clearly highlight the extent of the stray fields penetrating the windings in the transformer window for the specific cases of window height to width ratios of 0.5, 2 and 8, respectively.

It can be observed from these figures that the percentage of winding areas subject to relatively high stray magnetic fields for the transformers having window height to width ratios of 0.5 and 8 shown in Figs. 4.8 and 4.10, respectively, is higher than that of the transformer having an analogous ratio of 2 as shown in Fig. 4.9. More detailed studies for the estimation of the optimum window height to width ratio range of power transformers have been previously carried out (refer, for instance, to [11]). For design purposes leading to the minimization of eddy current losses in the transformer windings, it is recommended to maintain the window height to width ratio in the range between 2 and 2.5.

Denoting the window height to width ratio by c_{hw}, Eqs. (4.29) and (4.30) corresponding to single and three-phase transformers, respectively, may hence be rewritten in the forms:

$$S_w = \frac{c_{hw}(1 + c_j)N_H ac_H}{H_W^2},\tag{4.32}$$

$$S_w = \frac{2c_{hw}(1 + c_j)N_H ac_H}{H_W^2}.\tag{4.33}$$

Hence, for single and three-phase transformers, respectively, (4.32) and (4.34) may be rearranged in the forms:

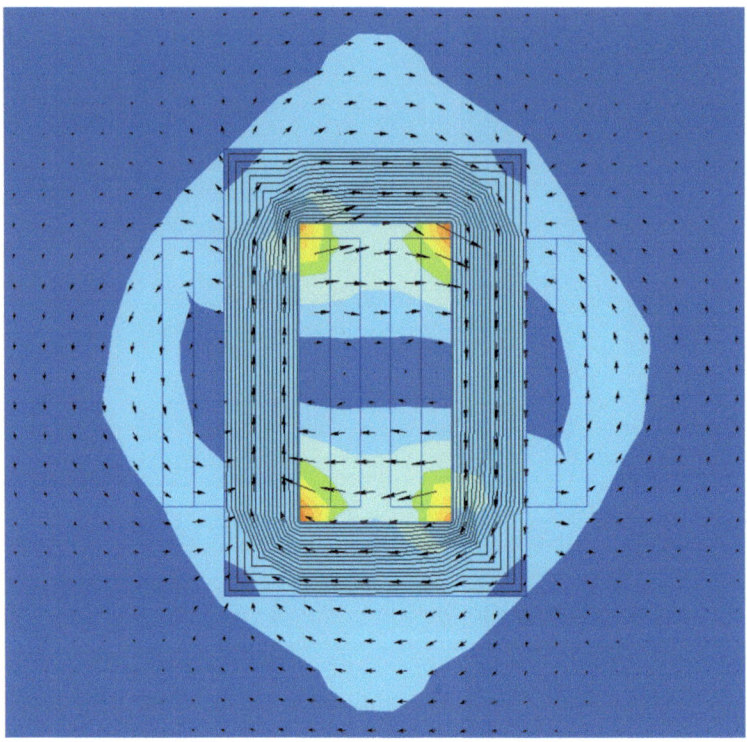

Fig. 4.9 Finite element analysis of the stray magnetic fields in a power transformer window having a window height to width ratio of 2

$$N_H ac_H = \frac{S_w H_W^2}{c_{hw}(1 + c_j)},\tag{4.34}$$

$$N_H ac_H = \frac{S_w H_W^2}{2c_{hw}(1 + c_j)}.\tag{4.35}$$

4.5   The Output Equation

It is well known that the rating of a single-phase power transformer may be simply calculated from the expression:

$$S = V_H I_H,\tag{4.36}$$

where I_H and V_H are the high-voltage current and voltage for a single-phase transformer, respectively.

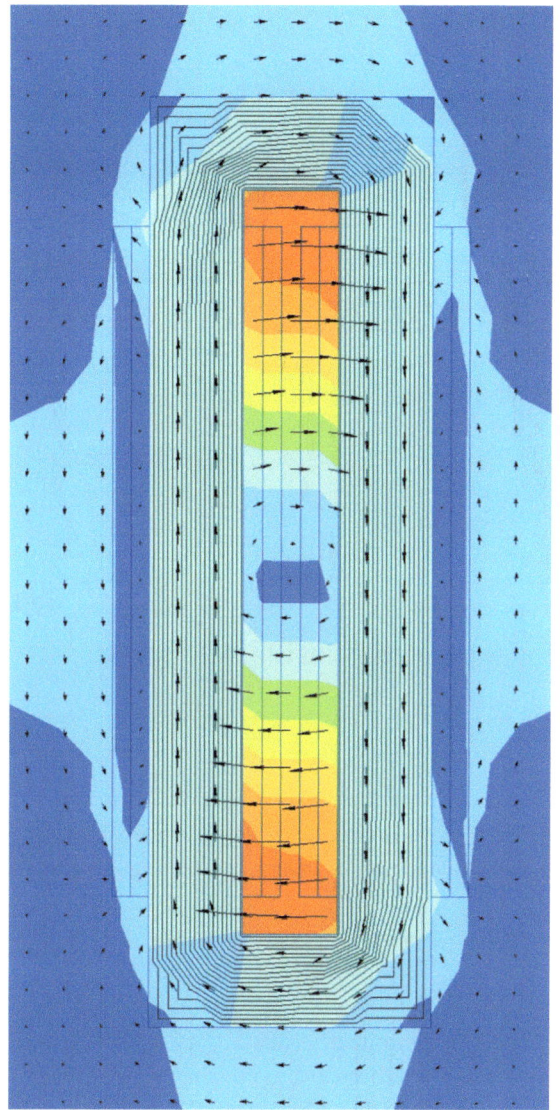

Fig. 4.10 Finite element analysis of the stray magnetic fields in a power transformer window having a window height to width ratio of 8

Considering the primary voltage side to be the higher voltage one and substituting (4.15) into (4.37), we get:

$$S = 4.44\left(0.95K_c\frac{\pi}{4}\right)fB_cD^2N_HI_H. \tag{4.37}$$

From (4.21):

$$S = 4.44\left(0.95K_c\frac{\pi}{4}\right)fB_cD^2J_HN_Hac_H. \tag{4.38}$$

Substituting (4.35) into (4.39), we get:

$$S = \left[\frac{4.44 S_w}{c_{hw}(1 + c_j)}\left(0.95 K_{cf}\frac{\pi}{4}\right)\right] J_H B_c D^2 H_W^2. \tag{4.39}$$

Likewise, the rating of a three-phase power transformer may be calculated from the expression:

$$S = 3 V_{phH} I_{phH}, \tag{4.40}$$

where I_{phH} and V_{phH} are the high-voltage phased current and voltage, respectively.

Once more, considering the primary voltage side to be the higher voltage one and substituting (4.17) into (4.41) we get:

$$S = 3\left[4.44\left(0.95 K_c\frac{\pi}{4}\right)\right] f B_c D^2 N_H I_{phH}. \tag{4.41}$$

From (4.23):

$$S = 3\left[4.44\left(0.95 K_c\frac{\pi}{4}\right)\right] f B_c D^2 J_H N_H a c_H. \tag{4.42}$$

Substituting (4.36) into (4.43), we get:

$$S = 3\left[\frac{4.44 S_w}{2 c_{hw}(1 + c_j)}\left(0.95 K_{cf}\frac{\pi}{4}\right)\right] J_H B_c D^2 H_W^2. \tag{4.43}$$

It can be inferred from Eqs. (4.39) and (4.43) that the rating of a single or three-phase power transformer, for a set of dimensional specifics, may be expressed in terms of its electric loading J_H, magnetic loading B_c, core diameter D and window height H_W.

4.6   Total Copper Losses

The overall copper losses P_{cu} of a power transformer may be categorized as the windings ohmic losses $P_{cu\text{-}ohmic}$, the windings eddy current losses $P_{cu\text{-}eddy}$, and the copper termination connection losses $P_{cu\text{-}con}$.

In the absence of detailed electromagnetic field analysis of a transformer, accurate estimation of the windings eddy current losses is not possible. Nevertheless, by adopting a reasonable transformer window height to width ratio as discussed in Sect. 4.4, the windings eddy current losses $P_{cu\text{-}eddy}$ may be reasonably assumed have a 15% ceiling of the windings ohmic losses $P_{cu\text{-}ohmic}$. Moreover, the copper termination connection losses $P_{cu\text{-}con}$ are also difficult to assess without detailed analysis that takes into account the material and shape configuration of the transformer terminals. They also may be reasonably assumed to have a 5% ceiling of the windings ohmic losses $P_{cu\text{-}ohmic}$.

Given that the design procedure should guarantee meeting the given specifications requirements which usually include the copper losses, the worst-case value of the transformer copper losses should be assumed in case no detailed computational software and/or analytical analysis are available. In this case, the total transformer copper losses may be given in the form:

$$P_{cu} = P_{cu-ohmic} + P_{cu-eddy} + P_{cu-con}, \tag{4.44}$$

$$P_{cu} = P_{cu-ohmic} + 0.15P_{cu-ohmic} + 0.05P_{cu-ohmic} = 1.2P_{cu-ohmic}. \tag{4.45}$$

From (4.46), the copper losses for the case of a single-phase power transformer may be expressed in the form:

$$P_{cu} \approx 1.2\left(I_H^2 \frac{\rho_{cu}N_H l_{mt}}{ac_H} + I_L^2 \frac{\rho_{cu}N_L l_{mt}}{ac_L}\right), \tag{4.46}$$

where ρ_{cu} is the resistivity of copper, while l_{mt} is the average turn length for both the high and low voltage windings.

From (4.21) and (4.22), (4.47) may be rewritten in the form:

$$P_{cu} \approx 1.2\rho_{cu}\left(J_H^2 N_H ac_H l_{mt} + J_L^2 N_L ac_L l_{mt}\right). \tag{4.47}$$

From (4.25) and (4.28), we obtain:

$$P_{cu} \approx 1.2\rho_{cu}\left(J_H^2 N_H ac_H l_{mt} + \frac{J_H^2}{c_j^2}c_j N_H ac_H l_{mt}\right). \tag{4.48}$$

Hence,

$$P_{cu} \approx 1.2J_H^2 \rho_{cu}\left(1 + \frac{1}{c_j}\right)N_H ac_H l_{mt}. \tag{4.49}$$

Denoting the high-voltage winding volume and low-voltage winding volume by Vol_{cu-H} and Vol_{cu-L}, respectively, (4.48) and (4.50) may be expressed in the form:

$$P_{cu} \approx 1.2\rho_{cu}\left(J_H^2 Vol_{cu-H} + J_L^2 Vol_{cu-L}\right) = 1.2J_H^2 \rho_{cu}\left(1 + \frac{1}{c_j}\right)Vol_{cu-H}. \tag{4.50}$$

For the case of a three-phase power transformer and referring to (4.46), the copper losses may be expressed in the form:

$$P_{cu} \approx 1.2\left(3I_{phH}^2 \frac{\rho_{cu}N_H l_{mt}}{ac_H} + 3I_{phL}^2 \frac{\rho_{cu}N_L l_{mt}}{ac_L}\right), \tag{4.51}$$

Once more from (4.23) and (4.24), (4.52) may be rewritten in the form:

$$P_{cu} \approx 1.2\rho_{cu}\left(3J_H^2 N_H ac_H l_{mt} + 3J_L^2 N_L ac_L l_{mt}\right). \tag{4.52}$$

From (4.25) and (4.28), we obtain:

$$P_{cu} \approx 1.2\rho_{cu}\left(3J_H^2 N_H ac_H l_{mt} + 3\frac{J_H^2}{c_j^2} c_j N_H ac_H l_{mt}\right). \tag{4.53}$$

Hence,

$$P_{cu} \approx 1.2 J_H^2 \rho_{cu}\left(1 + \frac{1}{c_j}\right) 3N_H ac_H l_{mt}. \tag{4.54}$$

In other words,

$$P_{cu} \approx 1.2\rho_{cu}\left(J_H^2 Vol_{cu-H} + J_L^2 Vol_{cu-L}\right) = 1.2 J_H^2 \rho_{cu}\left(1 + \frac{1}{c_j}\right) Vol_{cu-H}. \tag{4.55}$$

It turns out that the average turn length l_{mt} may be reasonably assessed in terms of the transformer window dimensions. Consider the single-phase transformer window shown in Fig. 4.7. Based on this figure, the average turn length may be given by:

$$l_{mt} = \pi(D + b_w + a_w). \tag{4.56}$$

As a rough approximation, the relation between the average winding width b_w may be assumed to be four times the windings interspacing a_w. In this case, (4.57) may be expressed as:

$$l_{mt} \approx \pi(D + \frac{5b_w}{4}). \tag{4.57}$$

Moreover, as can be seen from Fig. 4.7, the window width may be expressed as:

$$W_W = 2b_w + 4a_w = 3b_w. \tag{4.58}$$

Thus,

$$b_w = \frac{W_W}{3}. \tag{4.59}$$

Substituting (4.60) into (4.58), we get:

$$l_{mt} \approx \pi\left(D + \frac{5W_w}{12}\right) = \pi\left(D + \frac{5H_w}{12c_{hw}}\right). \tag{4.60}$$

From (4.50) and (4.61), the total copper losses for a single-phase power transformer may be written in the form:

$$P_{cu} \approx 1.2 J_H^2 \rho_{cu} \left[\pi \left(1 + \frac{1}{c_j} \right) \left(D + \frac{5H_w}{12c_{hw}} \right) \right] N_H a c_H . \tag{4.61}$$

Substituting (4.35) into (4.62), we get:

$$P_{cu} \approx 1.2 J_H^2 \rho_{cu} \left[\pi \left(1 + \frac{1}{c_j} \right) \left(D + \frac{5H_w}{12c_{hw}} \right) \right] \frac{S_w H_W^2}{c_{hw} (1 + c_j)} , \tag{4.62}$$

$$P_{cu} \approx 1.2 \rho_{cu} \left[\pi \left(\frac{S_w}{c_j c_{hw}} \right) \right] J_H^2 H_W^2 \left(D + \frac{5H_w}{12c_{hw}} \right) , \tag{4.63}$$

For the case of a three-phase power transformer and referring to the transformer window shown in Fig. 4.7, the average turn length l_{mt} may be given by:

$$l_{mt} = \pi (D + 2b_w + a_w) . \tag{4.64}$$

Once more, assuming the average winding width b_w to be four times the windings interspacing a_w, (4.65) may be rewritten in the form:

$$l_{mt} \approx \pi \left(D + \frac{9b_w}{4} \right) . \tag{4.65}$$

The window width as deduced from Fig. 4.7 may also be expressed as:

$$W_W = 4b_w + 4a_w = 5b_w . \tag{4.66}$$

Thus,

$$b_w = \frac{W_W}{5} . \tag{4.67}$$

Substituting (4.68) into (4.66), we get:

$$l_{mt} \approx \pi \left(D + \frac{9W_w}{20} \right) = \pi \left(D + \frac{9H_w}{20c_{hw}} \right) . \tag{4.68}$$

Hence, from (4.55) and (4.69), the total copper losses for a three-phase power transformer may be written in the form:

$$P_{cu} \approx 3.6 J_H^2 \rho_{cu} \left[\pi \left(1 + \frac{1}{c_j} \right) \left(D + \frac{9H_w}{20c_{hw}} \right) \right] N_H a c_H . \tag{4.69}$$

Substituting (4.36) into (4.70), we get:

$$P_{cu} \approx 1.8 \rho_{cu} \left[\pi \left(\frac{S_w}{c_j c_{hw}} \right) \right] J_H^2 H_W^2 \left(D + \frac{9H_w}{20c_{hw}} \right) . \tag{4.70}$$

As can be seen from expressions (4.63) and (4.70), the total copper losses of a single or three-phase power transformer, for a set of dimensional specifics, may be expressed in terms of its electric loading J_H, core diameter D and the window height H_W.

4.7 Total No Load Losses

The overall no-load losses P_{NL} may be regarded as a superposition of the core losses P_{Fe} and the difficult-to-account-for stray losses P_{stray}. Similar to the case of the windings eddy current losses, accurate estimation of the stray losses requires massive computational resources and high expertise that involve complex models of coupled magnetic, thermal and mechanical factors.

It should be pointed out that the stray losses P_{stray} may, for example, be due to additional iron losses resulting from the uneven distribution of the flux density in the core, losses in steel surroundings such as mechanical winding clamps and steel tanks, and mechanical losses due to the magnetostriction phenomenon (please refer, for example, to [12–18]).

In absence of the analytical capability to deduce the stray losses and, once more, to guarantee meeting the required maximum no-load losses, a worst-case scenario is assumed as follows:

$$P_{NL} = P_{Fe} + P_{stray} \approx 1.3 P_{Fe}. \tag{4.71}$$

Denoting the specific iron loss per Kg for a specific frequency and maximum flux density by $W_{Fe}(B_c, f)$, (4.72) may be rewritten in the form:

$$P_{NL} \approx 1.3 W_{Fe}(B_c, f) \delta_{Fe} Vol_{Fe}, \tag{4.72}$$

where δ_{Fe} is the core laminations density and Vol_{Fe} is the core total volume.

It is worth mentioning that the specific iron loss per Kg for a specific frequency and maximum flux density $W_{Fe}(B_c, f)$ is usually available in the specifications data sheet for the particular steel laminations used in the power transformer construction (see, for instance, [19]). As suggested by (3.37), $W_{Fe}(B_c, f)$ may be approximated for the operating supply frequency, using a curve fitting algorithm, in the form:

$$W_{Fe} \approx c_{Fe} B_c^2. \tag{4.73}$$

Substituting (4.74) into (4.73), we get:

$$P_{NL} \approx (1.3 \delta_{Fe} c_{Fe}) B_c^2 Vol_{Fe}. \tag{4.74}$$

For a single-phase core type transformer, the core volume Vol_{Fe} may be deduced from:

$$Vol_{Fe} = \left(0.95K_c\frac{\pi}{4}D^2\right)[2H_W + 2(W_W + 2D)]. \tag{4.75}$$

Substituting (4.76) into (4.75), we get:

$$P_{NL} \approx \left(1.3\delta_{Fe}c_{Fe}0.95K_c\frac{\pi}{4}\right)B_c^2D^2\left[2H_W + 2\left(\frac{H_W}{c_{hw}} + 2D\right)\right], \tag{4.76}$$

$$P_{NL} \approx \left(1.3\delta_{Fe}c_{Fe}0.95K_c\frac{\pi}{4}\right)B_c^2D^2\left[2H_W\left(1 + \frac{1}{c_{hw}}\right) + 4D\right]. \tag{4.77}$$

On the other hand, for a three-phase core type transformer the volume Vol_{Fe} may be deduced from:

$$Vol_{Fe} = \left(0.95K_c\frac{\pi}{4}D^2\right)[3H_W + 2(2W_W + 3D)]. \tag{4.78}$$

Substituting (4.79) into (4.75), we get:

$$P_{NL} \approx \left(1.3\delta_{Fe}c_{Fe}0.95K_c\frac{\pi}{4}\right)B_c^2D^2[3H_W + 2(2W_W + 3D)], \tag{4.79}$$

$$P_{NL} \approx \left(1.3\delta_{Fe}c_{Fe}0.95K_c\frac{\pi}{4}\right)B_c^2D^2\left[H_W\left(3 + \frac{4}{c_{hw}}\right) + 6D\right]. \tag{4.80}$$

As can be seen from (4.78) and (4.81), the total no-load losses of a single or three-phase power transformer, for a set of dimensional specifics, may be expressed in terms of its magnetic loading B_c, core diameter D and the window height H_W.

4.8 Leakage Reactance

With reference to Fig. 4.7, the power transformer leakage reactance referred to the high-voltage side X_{Heq} may be computed using the following two expressions for single and three-phase transformers, respectively (see [7] and [19]):

$$X_{Heq} = 2\pi f \mu_o N_H^2 \frac{l_{mt}}{l_{WH}}\left(a_w + \frac{2b_w}{6}\right), \tag{4.81}$$

$$X_{Heq} = 2\pi f \mu_o N_H^2 \frac{l_{mt}}{l_{WH}}\left(a_w + \frac{2b_w}{3}\right). \tag{4.82}$$

Practically, the average winding height l_{WH} may be approximated according to:

$$l_{WH} \approx 0.9H_W. \tag{4.83}$$

Substituting (4.61) and (4.84) into (4.82) and assuming again that the average winding width b_w to be four times the windings interspacing a_w, the leakage reactance referred to the high-voltage side of a single-phase power transformer may be expressed in the form:

$$X_{Heq} = \frac{2\pi f \mu_o N_H^2}{0.9H_W} \pi \left(D + \frac{5H_w}{12c_{hw}}\right)\left(\frac{b_w}{4} + \frac{2b_w}{6}\right), \tag{4.84}$$

$$X_{Heq} = \frac{2\pi f \mu_o N_H^2}{0.9H_W} \pi \left(D + \frac{5H_w}{12c_{hw}}\right)\left(\frac{7b_w}{12}\right). \tag{4.85}$$

Substituting (4.60) into (4.86), we obtain:

$$X_{Heq} = \frac{2\pi f \mu_o N_H^2}{0.9H_W} \pi \left(D + \frac{5H_w}{12c_{hw}}\right)\left(\frac{7W_w}{36}\right), \tag{4.86}$$

$$X_{Heq} = \left[\frac{14\pi^2 f \mu_o}{0.9(36)c_{hw}}\right] N_H^2 \left(D + \frac{5H_w}{12c_{hw}}\right). \tag{4.87}$$

Substituting (4.35) into (4.88), we get:

$$X_{Heq} = \left[\frac{14\pi^2 f \mu_o}{0.9(36)c_{hw}}\right]\left[\frac{S_w H_W^2}{a c_H C_{hw}(1 + c_j)}\right]^2 \left(D + \frac{5H_w}{12c_{hw}}\right). \tag{4.88}$$

From (4.21) and rearranging (4.89), we end up with the following expression for the single-phase power transformer leakage reactance referred to the high-voltage side:

$$X_{Heq} = \left[\frac{14\pi^2 f \mu_o S_w^2}{0.9(36)c_{hw}^3(1 + c_j)^2 I_H^2}\right] J_H^2 H_W^4 \left(D + \frac{5H_w}{12c_{hw}}\right). \tag{4.89}$$

A similar expression may be deduced for three-phase power transformers. Substituting (4.69) and (4.84) into (4.83) and assuming again that the average winding width b_w is four times the windings interspacing a_w, the leakage reactance referred to the high-voltage side of a three-phase power transformer may be expressed in the form:

$$X_{Heq} = \frac{2\pi f \mu_o N_H^2}{0.9H_W} \pi \left(D + \frac{9H_w}{20c_{hw}}\right)\left(\frac{11b_w}{12}\right), \tag{4.90}$$

Substituting (4.68) into (4.91), we obtain:

$$X_{Heq} = \frac{2\pi f \mu_o N_H^2}{0.9H_W} \pi \left(D + \frac{9H_w}{20c_{hw}}\right)\left(\frac{11W_w}{60}\right), \tag{4.91}$$

$$X_{Heq} = \left[\frac{22\pi^2 f \mu_o}{0.9(60)c_{hw}}\right] N_H^2 \left(D + \frac{9H_w}{20c_{hw}}\right). \tag{4.92}$$

Substituting (4.36) into (4.93), we get:

$$X_{Heq} = \left[\frac{22\pi^2 f \mu_o}{0.9(60)c_{hw}} \right] \left[\frac{S_w H_W^2}{2ac_H c_{hw}(1+c_j)} \right]^2 \left(D + \frac{9H_w}{20c_{hw}} \right). \tag{4.93}$$

From (4.23) and rearranging (4.94) we end up with the following expression for the three-phase power transformer leakage reactance referred to the high-voltage side:

$$X_{Heq} = \left[\frac{11\pi^2 f \mu_o S_w^2}{1.8(60)c_{hw}^3(1+c_j)^2 I_{phH}^2} \right] J_H^2 H_W^4 \left(D + \frac{9H_w}{20c_{hw}} \right). \tag{4.94}$$

Equations (4.89) and (4.94) clearly suggest that, for the case of specific design requirements, the single and three-phase power transformer leakage reactance referred to the high-voltage side may be computed in terms of the electric loading J, core diameter D and the window height H_W.

4.9   Magnetizing Reactance

Knowing the power transformer core weight, its magnetizing reactance may be deduced from the specifications sheet of the particular steel laminations used to construct this transformer [20]. More specifically, specifications sheet include the magnetizing root mean square volt ampere per Kg for any specific maximum flux density and supply frequency $VA_{mag}(B_c, f)$ (see, for instance, [19]).

For single-phase transformers and referring to (4.76), the magnetizing current I_{mH} referred to the high-voltage side may be deduced from:

$$I_{mH} = \frac{VA_{mag}(B_c, f)}{V_H} \left(0.95 K_c \frac{\pi}{4} D^2 \right) \left[2H_W \left(1 + \frac{1}{c_{hw}} \right) + 4D \right]. \tag{4.95}$$

Hence, the magnetizing reactance referred to the high-voltage side of a single-phase power transformer may be computed from:

$$X_{mH} = \frac{V_H}{I_{mH}}. \tag{4.96}$$

On the other hand, for three-phase transformers and referring to (4.79), the magnetizing phase current I_{phmH} referred to the high-voltage side may be deduced from:

$$I_{phmH} = \frac{VA_{mag}(B_c, f)}{3V_{phH}} \left(0.95 K_c \frac{\pi}{4} D^2 \right) \left[H_W \left(3 + \frac{4}{c_{hw}} \right) + 6D \right]. \tag{4.97}$$

The magnetizing reactance per phase referred to the high-voltage side of a three-phase power transformer may thus be computed from:

$$X_{mH} = \frac{V_{phH}}{I_{phmH}}.$$

(4.98)

References

1. Constantin, D., Nicolae, P. M., & Nitu, C. M. (2013). 3D Finite element analysis of a three phase power transformer. In *Eurocon 2013* (pp. 1548–1552). IEEE.
2. Heathcote, M. J. (1998). *The J & P transformer book: A practical technology of the power transformer*. Newnes.
3. Adly, A. A. (2017). A specifications-oriented initial design methodology for power transformers. *Energy Systems, 8*, 285–296.
4. Adly, A. A., & Abd-El-Hafiz, S. K. (2015). A performance-oriented power transformer design methodology using multi-objective evolutionary optimization. *Journal of Advanced Research, 6*(3), 417–423.
5. Gerstl, A., & Karisch, S. E. (1997). Cost optimization for the slitting of core laminations for power transformers. *Annals of Operations Research, 69*, 157–169.
6. Adly, A. A. (2021). Computer-aided transformer design capacity building. *Transformers Magazine, 8*(1), 72–77.
7. Sawhney, A. K. (1999). *A course in electrical machine design, dhanpat rai & co.* Publications.
8. da Silva, J. R., Paganoto, P. S., Graeff, R., da Rocha, C. M., Bernartt, M. L., & dos Santos, C. W. (2020). Analysis of methods eddy current loss estimation in power transformer windings with multiphysical consideration (electromagnetic and fluid dynamic). In *2020 IEEE 19th biennial conference on electromagnetic field computation (CEFC)* (pp. 1–4). IEEE.
9. Thango, B. A., Jordaan, J. A., & Nnachi, A. F. (2021). A novel approach for evaluating eddy current loss in wind turbine generator step-up transformers. *Advances in Science, Technology and Engineering Systems Journal, 6*, 488–498.
10. Tera Analysis Ltd. (2021). QuickField finite element analysis system. Version 6.6 User's Guide. Available online: https://quickfield.com/downloads/quickfield_manual.pdf. Accessed on 15 Nov 2024.
11. Saleh, A., Adly, A., Fawzi, T., Omar, A., & El-Debeiky, S. (2002). Estimation and minimization techniques of eddy current losses in transformer windings. In *Proceedings of the CIGRE conference, Paris, France* (pp. 1–6).
12. Saleh, A., Omar, A., Amin, A., Adly, A., Fawzi, T., & El-Debeiky, S. (2004). Estimation and minimization techniques of transformer tank losses. In *Proceedings of the CIGRE conference, Paris, France* (pp. 1–6).
13. Adly, A. A., & Abd-El-Hafiz, S. K. (2009). Utilizing particle swarm optimization in the field computation of nonlinear media subject to mechanical stress. *Journal of Applied Physics, 105*(7).
14. Adly, A. A., & Abd-El-Hafiz, S. K. (2003). Identification and testing of an efficient Hopfield neural network magnetostriction model. *Journal of magnetism and magnetic materials, 263*(3), 301–306.

15. Adly, A. A., & Abd-El-Hafiz, S. K. (2008). Implementation of magnetostriction Preisach-type models using orthogonally coupled hysteresis operators. *Physica B: Condensed Matter, 403*(2–3), 425–427.
16. Adly, A. A., & Abd-El-Hafiz, S. K. (2016). Simulation of magneto-elastic materials using a novel vector hysteresis model. In *2016 13th international conference on electrical engineering/electronics, computer, telecommunications and information technology (ECTI-CON)* (pp. 1–6). IEEE.
17. Adly, A. A., & Abd-El-Hafiz, S. K. (2018). Construction of a magnetostrictive hysteresis operator using a tripod-like primitive hopfield neural network. *AIP Advances, 8*(5).
18. Adly, A. A., & Abd-El-Hafiz, S. K. (2020). An efficient vector hysteresis model for unidirectional magneto-elastic interactions. *IEEE Transactions on Magnetics, 57*(2), 1–5.
19. Steel, A. K. (2013). TRAN-COR H grain oriented electrical steels, product data bulletin. *Ohio, USA, 1.*
20. Hernandez, I. A., Cañedo, J. M., Olivares-Galvan, J. C., & Betancourt, E. (2015). Novel technique to compute the leakage reactance of three-phase power transformers. *IEEE Transactions on Power Delivery, 31*(2), 437–444.

Fulfilling Required Design Specifications and the Role of Electromagnetic Field Computations in Achieving Accurate Design

It has been shown in Chap. 4 that designing a power transformer to meet a required set of specifications may be analytically reduced to the determination of the optimum electric loading, magnetic loading, core diameter and window height that would lead to the fulfilment of the transformer volt-ampere (VA) rating, maximum copper losses, maximum no-load losses and leakage reactance. It has also been shown that, through the analytical reasoning, all other design parameters may be deduced from the abovementioned optimum design values, materials specifications and design assumptions within limited pre-identified ranges. In this chapter, identification of the design problem is further elaborated and the impact of utilizing electromagnetic field calculations in the accurate design process and/or cost minimization is discussed.

5.1 Fulfilling Required Design Specifications

For the case of a single-phase power transformer, it has been shown in Chap. 4 that the design problem mainly reduces to the identification of J_H, B_c, D and H_W that satisfy the derived Eqs. (4.39), (4.63), (4.77) and (4.89). Using those four equations, the following four variables can be defined:

$$G_{1-1} = \left[\frac{4.44 S_w}{c_{hw}(1 + c_j)} \left(0.95 K_c f \frac{\pi}{4} \right) \right],\tag{5.1}$$

$$G_{1-2} = 1.2 \rho_{cu} \left[\pi \left(\frac{S_w}{c_j c_{hw}} \right) \right],\tag{5.2}$$

A. Adly and S. Abd-El-Hafiz, *Unconventional Performance Oriented Power Transformers Design Methodologies*, Synthesis Lectures on Electrical Engineering, https://doi.org/10.1007/978-3-031-85221-3_5

$$G_{1-3} = \left(1.38_{Fe}c_{Fe}0.95K_c\frac{\pi}{4}\right),$$
(5.3)

$$G_{1-4} = \left[\frac{14\pi^2 f \mu_o S_w^2}{0.9(36)c_{hw}^3 (1+c_j)^2 I_H^2}\right].$$
(5.4)

Thus, Eqs. (4.39), (4.63), (4.77) and (4.89) may be rewritten in the form:

$$S = G_{1-1}J_H B_c D^2 H_W^2,$$
(5.5)

$$P_{cu} = G_{1-2}J_H^2 H_W^2\left(D+\frac{5H_w}{12c_{hw}}\right),$$
(5.6)

$$P_{NL} = G_{1-3}B_c^2 D^2\left[2H_W\left(1+\frac{1}{c_{hw}}\right)+4D\right],$$
(5.7)

$$X_{Heq} = G_{1-4}J_H^2 H_W^4\left(D+\frac{5H_w}{12c_{hw}}\right).$$
(5.8)

It has also been shown in Chap. 4 that the design problem of a three-phase power transformer may be reduced to the identification of J_H, B_c, D and H_W that satisfy the derived Eqs. (4.43), (4.70), (4.80) and (4.94). Using those four equations, the following four variables can also be defined:

$$G_{3-1} = 3\left[\frac{4.44S_w}{2c_{hw}(1+c_j)}\left(0.95K_c f\frac{\pi}{4}\right)\right],$$
(5.9)

$$G_{3-2} = 1.8\rho_{cu}\left[\pi\left(\frac{S_w}{c_j c_{hw}}\right)\right],$$
(5.10)

$$G_{3-3} = \left(1.38_{Fe}c_{Fe}0.95K_c\frac{\pi}{4}\right),$$
(5.11)

$$G_{3-4} = \left[\frac{11\pi^2 f \mu_o S_w^2}{1.8(60)c_{hw}^3 (1+c_j)^2 I_{phH}^2}\right].$$
(5.12)

Consequently, Eqs. (4.43), (4.70), (4.80) and (4.94) may be rewritten in the form:

$$S = G_{3-1}J_H B_c D^2 H_W^2,$$
(5.13)

$$P_{cu} = G_{3-2}J_H^2 H_W^2\left(D+\frac{9H_w}{20c_{hw}}\right),$$
(5.14)

$$P_{NL} = G_{3-3}B_c^2 D^2\left[H_W\left(3+\frac{4}{c_{hw}}\right)+6D\right],$$
(5.15)

$$X_{Heq} = G_{3-4} J_H^2 H_W^4 \left(D + \frac{9H_w}{20c_{hw}} \right).$$ (5.16)

It should be pointed out that all variables comprising G_{1-1}, G_{1-2}, G_{1-3}, G_{1-4}, G_{3-1}, G_{3-2}, G_{3-3}, and G_{3-4} may be evaluated as elaborated in Table 5.1.

A very important aspect of the introduced design methodology of this book is the reduction of the transformer design problem to the solution of four simultaneous nonlinear equations. More specifically, the power transformer design problem is reduced to the solution of Eqs. (5.9)–(5.12) for single-phase transformers and Eqs. (5.13)–(5.16) for three-phase transformers [1]. Different approaches to solve these equations will be covered in detail in the following two chapters. Although there are assumptions involved in this design methodology, these assumptions are in line with reported literature ranges as well as actual design cases. Eventually, adopting the proposed methodology should lead to a reasonable design that meets all required specifications. The design may be utilized as is or taken as an initial design for further design ramification to minimize the copper and/ or core material used, thus reducing the transformer production cost.

Table 5.1 Assessment of G_{1-1}, G_{1-2}, G_{1-3}, G_{1-4}, G_{3-1}, G_{3-2}, G_{3-3}, and G_{3-4} variables used in power transformer design

Variable	Assessment
Window space factor S_w	Reasonable assumption taking (4.31) as a guide
Core laminations stacking fill factor K_c	Dependent on the number of laminations steps used to form the core cross-section. Practically, more than 10 steps are used. Depending on the number of steps, it can reach a value of around 0.95
Frequency f	A design requirement
Window height to width ratio c_{hw}	A designer pick, optimally in the range between 2 and 2.5 (refer to Sect. 4.4)
Ratio between high to low current density c_j	A designer pick in the range between 1 and 1.3 (refer to Sect. 4.3)
Copper winding resistivity ρ_{cu}	May be found from the winding conductors data sheets. A reasonable assumption is 0.021 micro Ω-m
Steel core laminations density δ_{Fe}	May be found from the core laminations data sheets. A reasonable assumption is 7650 kg/m^3
Steel core laminations specific loss curve fitting coefficient c_{Fe}	May be deduced by curve fitting specific loss data given in the core laminations data sheets as per (4.73)
High voltage phase current I_{phH}	May be deduced from the required specifications

5.2   Role of Electromagnetic Field Computations in Achieving Accurate Design

Electromagnetic field computations can support achieving an accurate power transformer design at a minimum construction cost. This is not always an option for a wide range of power transformer manufacturers due to the high cost of relevant software packages and the need for high computational computer resources.

Utilizing electromagnetic field computation tools can be partial or comprehensive. Examples of the partial utilization of electromagnetic field computation is the accurate deduction of the eddy current losses in transformer windings as well as the copper termination losses. As suggested in (4.45), those losses were assumed to be no more than 20% of the ohmic copper losses. Accurate assessment of those losses might result in a lesser percentage, the fact that can be translated to a possible increase margin for the electric loading and, consequently, a reduction in the windings copper volume. Another example of the partial utilization of electromagnetic field computations is the accurate assessment of the no-load stray losses that were assumed to be no more than 30% of the core losses as suggested in (4.72). Bold examples of such losses are the tank eddy current losses and core losses due to uneven distribution of the core flux density. Accurate assessment of those losses might reveal a lesser percentage and this could offer a margin for slight increase in the design magnetic loading. Consequently, a reduction of core volume might be possible (refer, for instance, to [2–4]).

More comprehensive utilization of the (costly and computational resource demanding) electromagnetic field analysis tools can lead to the accurate determination of the power transformer performance (refer, for instance, to [5] and [6]). However, dimensional and material design details have to be known a-priori to perform the computation. In other words, while electromagnetic field computation tools can accurately predict the transformer performance (whether it be losses or reactance), initial design details should be available in advance. The more an initial design is close to being optimum, the smaller number of time-consuming electromagnetic field analysis sessions will be required. This further highlights the importance of having a reasonably accurate analytical design approach, which is the main focus of this book.

References

1. Adly, A. A. (2017). A specifications-oriented initial design methodology for power transformers. *Energy Systems, 8*, 285–296.
2. Saleh, A., Adly, A., Fawzi, T., Omar, A., & El-Debeiky, S. (2002). Estimation and minimization techniques of eddy current losses in transformer windings. In *Proceedings of the CIGRE conference, Paris, France* (pp. 1–6).
3. Thango, B. A., Jordaan, J. A., & Nnachi, A. F. (2021). Analysis of stray losses in transformers using finite element method modelling. In *2021 IEEE PES/IAS power Africa* (pp. 1–5). IEEE.

4. Thango, B. A., & Bokoro, P. N. (2022). Stray load loss valuation in electrical transformers: a review. *Energies, 15*(7), 2333.
5. Tsili, M. A., Kladas, A. G., & Georgilakis, P. S. (2008). Computer aided analysis and design of power transformers. *Computers in Industry, 59*(4), 338–350.
6. Jimenez-Mondragon, V. M., Escarela-Perez, R., Melgoza, E., Arjona, M. A., & Olivares-Galvan, J. C. (2017). Quasi-3-D finite-element modeling of a power transformer. *IEEE Transactions on Magnetics, 53*(6), 1–4.

Conventional Computer-Aided-Design (CAD) Approaches

A detailed power transformer design analytical methodology has been introduced in the previous chapters of the book. Within this methodology, the design problem was reduced to the determination of the optimum electric loading, magnetic loading, core diameter and window height that would lead to the fulfilment of the required power transformer lead specifications (i.e., the VA rating, maximum copper losses, maximum no-load losses and leakage reactance). This chapter is dedicated to presenting alternative approaches to solve the reduced design equations outlined in Chap. 5. It should be reiterated here that once the optimum electric loading, magnetic loading, core diameter and window height are determined, all remaining design details may be inferred as per the detailed derivations presented in Chap. 4.

6.1   Analytical Design Approach

The design equations presented in Chap. 5 can be solved analytically for single-phase and three-phase transformers as will be described in this section, respectively.

Single-Phase Transformers: Inspection of the relevant design Eqs. (5.5)–(5.8) suggest that an optimum window height H_W may be deduced by dividing Eq. (5.8) by (5.6) as follows:

$$\frac{X_{Heq}}{P_{cu}} = \frac{G_{1-4}J_H^2 H_W^4\left(D + \frac{5H_w}{12c_{hw}}\right)}{G_{1-2}J_H^2 H_W^2\left(D + \frac{5H_w}{12c_{hw}}\right)} = \frac{G_{1-4}H_W^2}{G_{1-2}}. \tag{6.1}$$

A. Adly and S. Abd-El-Hafiz, *Unconventional Performance Oriented Power Transformers Design Methodologies*, Synthesis Lectures on Electrical Engineering, https://doi.org/10.1007/978-3-031-85221-3_6

Hence,

$$H_W = \sqrt{\frac{G_{1-2}X_{Heq}}{G_{1-4}P_{cu}}}. \tag{6.2}$$

From (5.6) and (5.7), we get:

$$J_H = \sqrt{\frac{P_{cu}}{G_{1-2}H_W^2\left(D + \frac{5H_w}{12c_{hw}}\right)}}, \tag{6.3}$$

$$B_c = \sqrt{\frac{P_{NL}}{G_{1-3}D^2\left[2H_W\left(1 + \frac{1}{c_{hw}}\right) + 4D\right]}}. \tag{6.4}$$

Knowing the value of H_W from (6.2) and substituting (6.3) and (6.4) into (5.5), we get:

$$S = G_{1-1}\sqrt{\frac{P_{cu}}{G_{1-2}H_W^2\left(D + \frac{5H_w}{12c_{hw}}\right)}}\sqrt{\frac{P_{NL}}{G_{1-3}D^2\left[2H_W\left(1 + \frac{1}{c_{hw}}\right) + 4D\right]}}D^2H_W^2, \tag{6.5}$$

$$D^2 = \frac{S}{G_{1-1}H_w^2}\sqrt{\frac{G_{1-2}H_w^2\left(D + \frac{5H_w}{12c_{hw}}\right)}{P_{cu}}\frac{G_{1-3}D^2\left[2H_W\left(1 + \frac{1}{c_{hw}}\right) + 4D\right]}{P_{NL}}}. \tag{6.6}$$

Hence,

$$D^2 = D\sqrt{\left(\frac{S}{G_{1-1}H_w^2}\right)^2\left(\frac{G_{1-2}G_{1-3}H_w^2}{P_{cu}P_{NL}}\right)\left(u_1D^2 + u_2D + u_3\right)}, \tag{6.7}$$

$$D^2 = D\sqrt{\left(\frac{S}{G_{1-1}}\right)^2\left(\frac{G_{1-2}G_{1-3}}{P_{cu}P_{NL}H_w^2}\right)\left(u_1D^2 + u_2D + u_3\right)}, \tag{6.8}$$

where

$$u_1 = 4, u_2 = \left[\frac{20}{12c_{hw}} + \left(2 + \frac{2}{c_{hw}}\right)\right]H_w, u_3 = \left(\frac{5}{12c_{hw}}\right)\left(2 + \frac{2}{c_{hw}}\right)H_w^2. \tag{6.9}$$

Moreover, define the following three variables:

$$v_1 = \left(\frac{S}{G_{1-1}}\right)^2\left(\frac{G_{1-2}G_{1-3}}{P_{cu}P_{NL}H_w^2}\right)u_1, \tag{6.10}$$

$$v_2 = \left(\frac{S}{G_{1-1}}\right)^2\left(\frac{G_{1-2}G_{1-3}}{P_{cu}P_{NL}H_w^2}\right)u_2, \tag{6.11}$$

$$v_3 = \left(\frac{S}{G_{1-1}}\right)^2 \left(\frac{G_{1-2}G_{1-3}}{P_{cu}P_{NL}H_w^2}\right)u_3. \tag{6.12}$$

Substituting (6.10)–(6.12) into (6.8) and squaring both sides, we get:

$$(1 - v_1)D^2 - v_2 D - v_3 = 0. \tag{6.13}$$

Hence, the core diameter may be computed from the expression:

$$D = \frac{v_2 + \sqrt{v_2^2 + 4(1 - v_1)v_3}}{2(1 - v_1)}. \tag{6.14}$$

Substituting (6.2) and (6.14) into (6.3) and (6.4), the remaining two unknowns J_H and B_c may be computed. Knowing H_W, D, J_H and B_c, all other design dimensions and details may be deduced using the reasoning presented in Chap. 4 (refer to [1] and [2]).

Example 6.1 It is required to design a single-phase power transformer that will be connected to a three-phase star/delta grid within a three identical single-phase transformer bank. The required specifications are given in Table 6.1.

Solution

For the steel material under consideration, the specific core loss may be fitted in accordance to (4.73) as shown in Fig. 6.1. In this figure, c_{Fe} values leading to these agreements were found to be equivalent to 0.2644 for the 0.23 mm H-0 type and 0.3234 for the 0.30 mm H-2 type, respectively.

Reasonably assumed design variables may be given by Table 6.2.

Based upon the required specifications and the reasonably assumed design parameters, G_{1-1}, G_{1-2}, G_{1-3}, and G_{1-4} were computed according to (5.1)–(5.4). From (6.2), (6.14),

Table 6.1 Required specifications for the single-phase power transformer of Example 6.1

Rating S	7.5 MVA
High voltage V_H	66 kV
Low voltage V_L	6.35 kV
Frequency f	50 Hz
Maximum copper losses P_{cu}	26 KW
Maximum no-load losses P_{NL}	6.9 KW
Leakage reactance referred to HV side X	60.87 Ω
Core laminations type	AK Steel Tran-Cor Grain Oriented steel 0.23 mm H-0 and 0.30 mm H-2 [3]

Fig. 6.1 Specific loss data for the core lamination materials under consideration and their analogous approximated fitted values

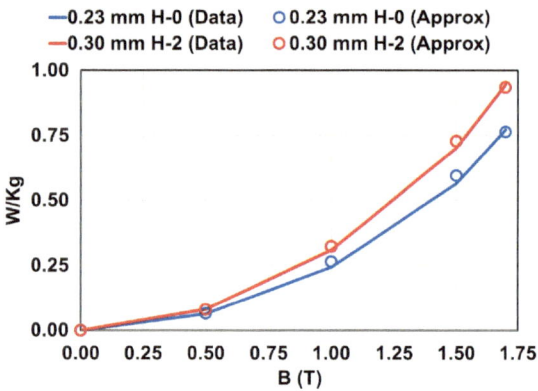

Table 6.2 Assumed additional design variables

Window space factor S_w	0.25
Core lamination stacking factor K_c	0.958
Window height to width c_{hw}	2.25
Current density ratio of the HV side to the LV side c_j	1.09
Copper winding resistivity ρ_{cu}	0.000000021
Steel core laminations density δ_{Fe}	7650
Steel core specific loss fitting parameter C_{Fe}	0.2644 for 0.23 mm H-0 0.3234 for 0.30 mm H-2
Copper winding density δ_{cu}	8933

(6.3) and (6.4), the optimum values of H_W, D, J_H and B_c that would satisfy the design requirements may be found. All other design details may then be computed based upon the detailed analyses and derivations given in Chap. 4. Computed results are given in Table 6.3 for both laminated core materials under consideration.

It is worth noting the effect of utilizing an inferior steel lamination brand having higher specific losses on the weight of steel and copper materials needed to fulfill the required design specifications. In other words, resorting to a lower cost steel brand might eventually lead to a more costly design. Hence, a feasibility study should always be carried out in accordance with the available materials and possible design details to achieve an optimum design at a minimum cost.

Three-Phase Transformers: A similar approach may be used for the design of three-phase power transformers. From Eqs. (5.13)–(5.16) it can be seen that the optimum window height H_W may be deduced by dividing Eq. (5.16) by (5.14) as follows:

Table 6.3 Computed design details using the analytical approach for the power transformer case of Example 6.1

Variable	Computed value	
	0.23 mm H-0	0.30 mm H-2
G_{1-1}	8.436128273	8.436128273
G_{1-2}	8.07015E−09	8.07015E−09
G_{1-3}	1879.512568	2298.919684
G_{1-4}	2.60656E−11	2.60656E−11
v_1	0.368784191	0.451077184
v_2	0.28489751	0.348471463
v_3	0.035750116	0.043727638
H_W (m)	0.851361559	0.851361559
D (m)	0.55364541	0.742163705
J_H (A/m^2)	2,499,792.817	2,222,559.15
B_c (T)	1.600746611	1.00192973
J_L (A/m^2)	2,293,387.906	2,039,045.092
W_W (m)	0.378382915	0.378382915
N_H	848	754
N_L	82	73
Wt_{Fe} (Kg)	7834.280297	16,349.00008
Wt_{Cu} (Kg)	1607.639275	2033.714491

$$\frac{X_{Heq}}{P_{cu}} = \frac{G_{3-4}J_H^2 H_W^4 \left(D + \frac{9H_w}{20c_{hw}}\right)}{G_{3-2}J_H^2 H_W^2 \left(D + \frac{9H_w}{20c_{hw}}\right)} = \frac{G_{3-4}H_W^2}{G_{3-2}}. \tag{6.15}$$

Thus,

$$H_W = \sqrt{\frac{G_{3-2}X_{Heq}}{G_{3-4}P_{cu}}}. \tag{6.16}$$

Moreover, from (5.14) and (5.15) we get:

$$J_H = \sqrt{\frac{P_{cu}}{G_{3-2}H_W^2 \left(D + \frac{9H_w}{20c_{hw}}\right)}}, \tag{6.17}$$

$$B_c = \sqrt{\frac{P_{NL}}{G_{3-3}D^2 \left[H_W \left(3 + \frac{4}{c_{hw}}\right) + 6D\right]}}. \tag{6.18}$$

Following the determination of H_W from (6.16) and substituting (6.17) and (6.18) into (5.13), we get:

$$S = G_{3-1}\sqrt{\frac{P_{cu}}{G_{3-2}H_W^2\left(D+\frac{9H_w}{20c_{hw}}\right)}}\sqrt{\frac{P_{NL}}{G_{3-3}D^2\left[H_W\left(3+\frac{4}{c_{hw}}\right)+6D\right]}}D^2H_W^2, \qquad (6.19)$$

$$D^2 = \frac{S}{G_{3-1}H_W^2}\sqrt{\frac{G_{3-2}H_w^2\left(D+\frac{9H_w}{20c_{hw}}\right)}{P_{cu}}\frac{G_{3-3}D^2\left[H_W\left(3+\frac{4}{c_{hw}}\right)+6D\right]}{P_{NL}}}. \qquad (6.20)$$

Hence,

$$D^2 = D\sqrt{\left(\frac{S}{G_{3-1}H_w^2}\right)^2\left(\frac{G_{3-2}G_{3-3}H_w^2}{P_{cu}P_{NL}}\right)(q_1D^2+q_2D+q_3)}, \qquad (6.21)$$

$$D^2 = D\sqrt{\left(\frac{S}{G_{3-1}}\right)^2\left(\frac{G_{3-2}G_{3-3}}{P_{cu}P_{NL}H_w^2}\right)(q_1D^2+q_2D+q_3)}, \qquad (6.22)$$

where,

$$q_1 = 6, q_2 = \left[\frac{54}{20c_{hw}}+\left(3+\frac{4}{c_{hw}}\right)\right]H_w, q_3 = \left(\frac{9}{20c_{hw}}\right)\left(3+\frac{4}{c_{hw}}\right)H_w^2. \qquad (6.23)$$

Moreover, define the following three variables:

$$\gamma_1 = \left(\frac{S}{G_{3-1}}\right)^2\left(\frac{G_{3-2}G_{3-3}}{P_{cu}P_{NL}H_w^2}\right)q_1, \qquad (6.24)$$

$$\gamma_2 = \left(\frac{S}{G_{3-1}}\right)^2\left(\frac{G_{3-2}G_{3-3}}{P_{cu}P_{NL}H_w^2}\right)q_2, \qquad (6.25)$$

$$\gamma_3 = \left(\frac{S}{G_{3-1}}\right)^2\left(\frac{G_{3-2}G_{3-3}}{P_{cu}P_{NL}H_w^2}\right)q_3. \qquad (6.26)$$

Substituting (6.24)–(6.26) into (6.22) and squaring both sides, we get:

$$(1-\gamma_1)D^2 - \gamma_2D - \gamma_3 = 0. \qquad (6.27)$$

The core diameter may thus be computed using the equation:

$$D = \frac{\gamma_2 + \sqrt{\gamma_2^2 + 4(1-\gamma_1)\gamma_3}}{2(1-\gamma_1)}. \qquad (6.28)$$

By substituting (6.16) and (6.28) into (6.17) and (6.18), the remaining two unknowns J_H and B_c may be computed. Once more, knowing H_W, D, J_H and B_c, all other design dimensions and details may be deduced using the reasoning presented in Chap. 4 (refer to [1] and [2]).

It should be stated here that a very important aspect of the deduced Eqs. (6.14) and (6.28) is the possibility of assessing whether a set of required specifications is achievable. Obviously, it is not achievable in the case when the radicand of the square root is negative. Unachievable specifications could be due to unrealistic low loss requirements for the utilized transformer core material, especially the steel laminations grade.

Example 6.2 It is required to design a three-phase delta/star ONAN/ONAF power transformer having the required set of specifications given in Table 6.4.

Solution

For the steel material under consideration, it has been shown in Fig. 6.1 that the specific core loss may be fitted in accordance to (4.73) by taking the value of c_{Fe} to be equivalent to 0.2644. By matching some of the design specifics for the actual power transformer under consideration, reasonable parameters were assumed as given in Table 6.5.

Based upon the required specifications and the reasonably assumed design parameters, G_{3-1}, G_{3-2}, G_{3-3}, and G_{3-4} were computed according to (5.9)–(5.12). From (6.16), (6.28), (6.17) and (6.18), the optimum values of H_W, D, J_H and B_c that would satisfy the design requirements may be found. All other design details may then be computed based upon the detailed analyses and derivations given in Chap. 4. The computed design details are given in Table 6.6.

It should be mentioned that the reported computed results are qualitatively and quantitatively in line with analogous design details of actual power transformers having the same specifications. In all cases, the design suggested by the proposed approach may also be regarded as a very good initial guess to be refined, if necessary, by more sophisticated computational tools.

Table 6.4 Required specifications for the three-phase power transformer of Example 6.2

Rating S	20 MV
High voltage phase value V_{phH}	66 kV
Low voltage phase value V_{phL}	6.35 kV
Frequency f	50 Hz
Maximum copper losses P_{cu}	80 KW
Maximum no-load losses P_{NL}	15 KW
Leakage reactance referred to HV side X	65.34 Ω
Core laminations type	AK Steel Tran-Cor Grain Oriented steel 0.23 mm H-0 [3]

Table 6.5 Reasonably assumed (or extracted from the actual transformer details) design parameters

Window space factor S_w	0.20
Core lamination stacking factor K_c	0.958
Window height to width c_{hw}	2.05
Current density ratio of the HV side to the LV side c_j	1.2
Copper winding resistivity ρ_{cu}	0.000000021
Steel core laminations density δ_{Fe}	7650
Steel core specific loss fitting parameter C_{Fe}	0.2644 for 0.23 mm H-0
Copper winding density δ_{cu}	8933

Table 6.6 Computed leading design details using the analytical approach for the three-phase power transformer case of Example 6.2

Variable	Computed value
G_{3-1}	10.5554483
G_{3-2}	9.65465E−09
G_{3-3}	1879.512568
G_{3-4}	5.93843E−12
γ_1	0.245304039
γ_2	0.295310969
γ_3	0.059003532
H_w (m)	1.15
D (m)	0.537
J_H (A/m^2)	2,810,762.43
B_c (T)	1.761
N_H	819
N_L	79
Wt_{Fe} (Kg)	14,071.73
Wt_{Cu} (Kg)	4307.44

Example 6.3 It is required to compare between the proposed approach design details and an actual three-phase delta/star ONAN/ONAF power transformer having the required set of specifications given in Table 6.7.

Solution

For the steel material under consideration, it has been shown in Fig. 6.1 that the specific core loss may be fitted in accordance to (4.73) by taking the value of c_{Fe} to be equivalent

Table 6.7 Required specifications for the three-phase power transformer of Example 6.3

Rating S	25 MV
High voltage phase value V_{phH}	66 kV
Low voltage phase value V_{phL}	6.35 kV
Frequency f	50 Hz
Maximum copper losses P_{cu}	99 KW
Maximum no-load losses P_{NL}	19.51 KW
Leakage reactance referred to HV side X	54.415 Ω
Core laminations type	AK Steel Tran-Cor Grain Oriented steel 0.23 mm H-0 [3]

Table 6.8 Reasonably assumed (or extracted from the actual transformer details) design parameters

Window space factor S_W	0.15
Core lamination stacking factor K_c	0.958
Window height to width c_{hw}	2.1
Current density ratio of the HV side to the LV side c_j	1.06
Copper winding resistivity ρ_{cu}	0.000000021
Steel core laminations density δ_{Fe}	7650
Steel core specific loss fitting parameter C_{Fe}	0.2644 for 0.23 mm H-0
Copper winding density δ_{cu}	8933

to 0.2644. By matching some of the design specifics for the actual power transformer under consideration, reasonable parameters were assumed as given in Table 6.8.

Based upon the required specifications and the reasonably assumed design parameters, G_{3-1}, G_{3-2}, G_{3-3}, and G_{3-4} were computed according to (5.9)–(5.12). From (6.16), (6.28), (6.17) and (6.18), the optimum values of H_W, D, J_H and B_c that would satisfy the design requirements may be found. All other design details may then be computed based upon the detailed analyses and derivations given in Chap. 4. A comparison between the computed and the actual results are given in Table 6.9.

It is clear from the comparison results shown in Table 6.9 that the proposed analytical approach can suggest valid design details that guarantee the fulfilment of a set of required three-phase power transformer specifications. Obviously, the design suggested by the proposed approach may also be regarded as a very good initial guess to be refined, if necessary, by more sophisticated computationally demanding electromagnetic field computation design packages.

Table 6.9 Comparison between actual and computed leading design details using the analytical approach for the three-phase power transformer case of Example 6.3

Variable	Actual value	Computed value	Error %
G_{3-1}	N/A	8.253306491	N/A
G_{3-2}	N/A	8.00217E$-$09	N/A
G_{3-3}	N/A	1879.512568	N/A
G_{3-4}	N/A	2.26824E$-$12	N/A
γ_1	N/A	0.221071389	N/A
γ_2	N/A	0.317619391	N/A
γ_3	N/A	0.075092361	N/A
H_w (m)	1.075	1.392520707	29.53
D (m)	0.54	0.575328903	6.54
J_H (A/m^2)	2,519,000.0	2,702,242.816	7.27
B_c (T)	1.737	1.746	0.52
N_H	798	719	9.90
N_L	83	69	16.87
Wt_{Fe} (Kg)	13,986	18,610.09	33.06
Wt_{Cu} (Kg)	5418	5094.35	5.97

6.2 Semi-Analytical Trial and Error Design Approach

A different trial-and-error approach to employ design Eqs. (5.5)–(5.8) and (5.13)–(5.16) for single-phase and three-phase transformers, respectively, may also be utilized.

Single-Phase Transformers: Consider the case of a single-phase power transformer where the optimum window height H_W is deduced according to (6.2). Assuming reasonable initial values for the electric and magnetic loading denoted by J_{H-i} and B_{c-i}, the core diameter D may be deduced from (5.5) as:

$$D = \sqrt{\frac{S}{G_{1-1}J_{H-i}B_{c-i}H_W^2}}. \tag{6.29}$$

Utilizing (5.6) and (5.7), the updated new values for the electric and magnetic loading J_{H-new} and B_{c-new}, respectively, may then be computed using:

$$J_{H-new} = \chi J_{H-i} + (1 - \chi)\sqrt{\frac{P_{cu}}{G_{1-2}H_W^2\left(D + \frac{5H_w}{12c_{hw}}\right)}}, \tag{6.30}$$

$$B_{c-new} = \chi B_{c-i} + (1 - \chi)\sqrt{\frac{P_{NL}}{G_{1-3}D^2\left[2H_W\left(1 + \frac{1}{c_{hw}}\right) + 4D\right]}}, \tag{6.31}$$

where χ is a number, whose value should be selected close to unity to guarantee convergence.

Defining a required accuracy criterion $\Delta\%$ for the electric and magnetic loading values, the convergence test reduces to checking the validity of the following inequalities:

$$\left| \frac{J_{H-new} - J_{H-i}}{J_{H-i}} \right| \times 100 \le \Delta\%, \tag{6.32}$$

$$\left| \frac{B_{c-new} - B_{c-i}}{B_{c-i}} \right| \times 100 \le \Delta\%. \tag{6.33}$$

If the accuracy criteria are satisfied, the computed electric and magnetic loadings are employed to deduce the remaining design details. If not, a new iteration is carried out until a maximum number of iterations is reached.

Three-Phase Transformers: Following the same reasoning for three-phase power transformers, where the optimum window height H_W is deduced according to (6.16), and assuming reasonable initial values for the electric and magnetic loading denoted by J_{H-i} and B_{c-i}, the core diameter D may be deduced from (5.13) as:

$$D = \sqrt{\frac{S}{G_{3-1} J_{H-i} B_{c-i} H_W^2}}. \tag{6.34}$$

From (5.14) and (5.15), the updated new values for the electric and magnetic loading J_{H-new} and B_{c-new}, respectively, may be computed from:

$$J_{H-new} = \chi J_{H-i} + (1 - \chi) \sqrt{\frac{P_{cu}}{G_{3-2} H_W^2 \left(D + \frac{9 H_w}{20 c_{hw}} \right)}}, \tag{6.35}$$

$$B_{c-new} = \chi B_{c-i} + (1 - \chi) \sqrt{\frac{P_{NL}}{G_{3-3} D^2 \left[H_W \left(3 + \frac{4}{c_{hw}} \right) + 6D \right]}}, \tag{6.36}$$

where, once more, χ is a number whose value should be selected close to unity to guarantee convergence.

By defining a required accuracy criterion $\Delta\%$ for the electric and magnetic loading values, the convergence test reduces to checking the validity of the inequalities given by (6.32) and (6.33), respectively. Obviously, if the accuracy criteria are satisfied, the computed electric and magnetic loadings are employed to deduce the remaining design details. If not, a new iteration is carried out until a maximum number of iterations is reached.

In summary, the flowchart of Fig. 6.2 depicts the presented trial-and-error semi-analytical approach for both single-phase and three-phase transformers.

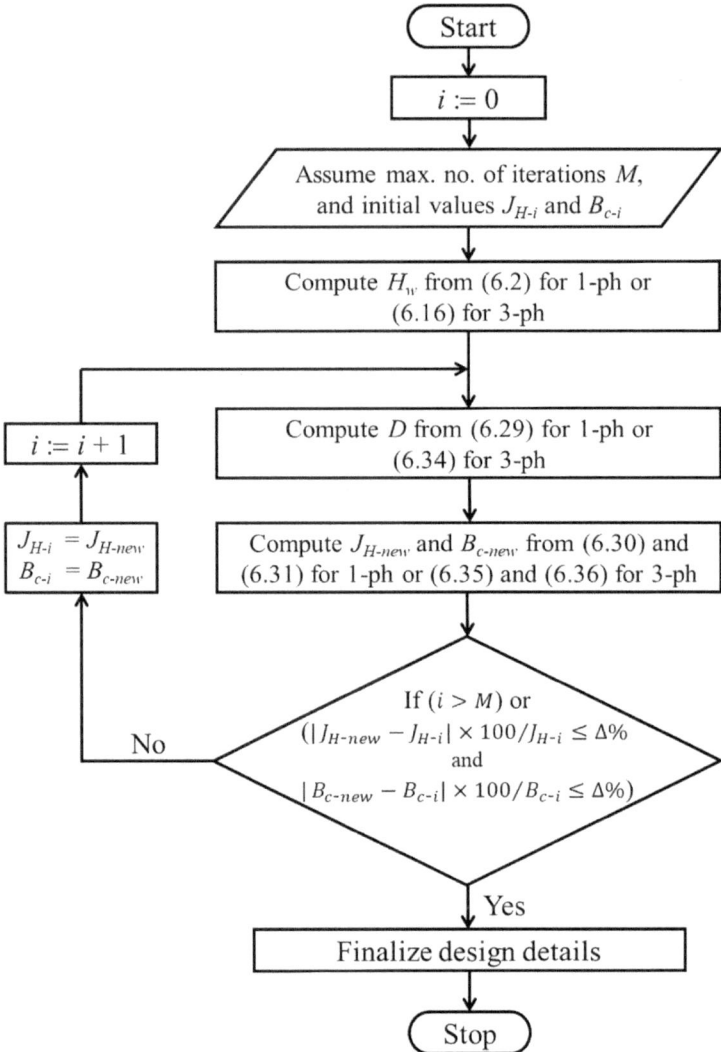

Fig. 6.2 A flowchart depicting the trial-and-error semi-analytical design approach for power transformers

Example 6.4 Using the semi-analytical trial-and-error approach, redesign the single-phase power transformer of Example 6.1 while considering the H-0 laminations material having the lower specific core losses.

Solution

For the steel material under consideration, it was demonstrated in Fig. 6.1 that the specific core loss c_{Fe} fitted in accordance with (4.73) may be assumed to be 0.2644 for the 0.23 mm H-0 type.

Taking the same reasonably assumed design values of Table 6.2, and based upon the required specifications and the reasonably assumed design parameters, G_{1-1}, G_{1-2}, G_{1-3}, and G_{1-4} were computed according to (5.1)–(5.4). From (6.2), the optimum value of H_W is computed. Applying the iterative approach given by (6.29)–(6.33) while taking χ to be 0.95 for a limited number of 300 iterations, J_H and B_c evolve as shown in Figs. 6.3 and 6.4, respectively.

After the trial-and-error iterations, estimates for D, J_H and B_c that would satisfy the design requirements should be found. All other design details may then be computed based upon the detailed analyses and derivations given in Chap. 4. Computed results are given in Table 6.10 for the laminated core material under consideration.

Fig. 6.3 Evolution of the HV side current density (corresponding to the 0.23 mm H-0 material) over the first 300 trial-and-error iterations for the single-phase transformer of Example 6.1

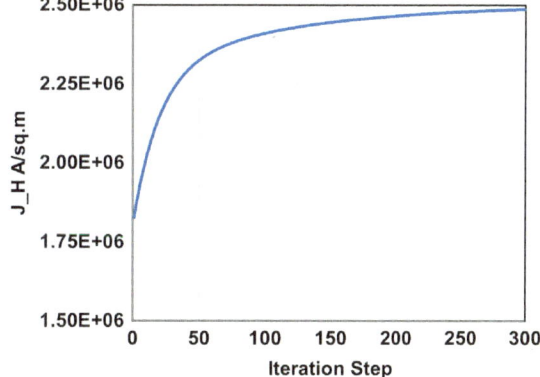

Fig. 6.4 Evolution of the core flux density (corresponding to the 0.23 mm H-0 material) over the first 300 trial-and-error iterations for the single-phase transformer of Example 6.1

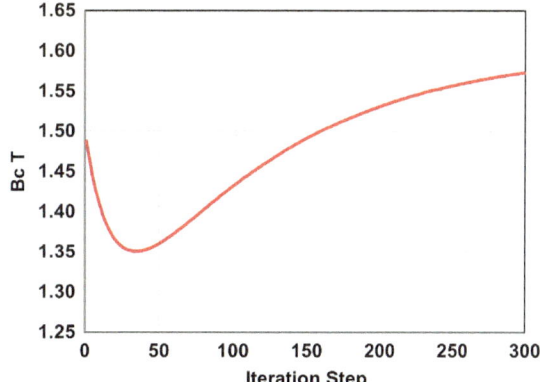

Table 6.10 Computed design details using the semi-analytical trial-and-error approach for the power transformer case of Example 6.1

Variable	Computed value for 0.23 mm H-0
G_{1-1}	8.436128273
G_{1-2}	8.07015E−09
G_{1-3}	1879.512568
G_{1-4}	2.60656E−11
H_w (m)	0.851361559
D (m)	0.560104473
J_H (A/m^2)	2,465,523.309
B_c (T)	1.532275767
J_L (A/m^2)	2,261,947.99
W_w (m)	0.378382915
N_H	836.0477134
N_L	80.44872873
Wt_{Fe} (Kg)	8062.463517
Wt_{Cu} (Kg)	1622.237574

It should be pointed out that after the 300 implemented iterations, the accuracy figures of J_H and B_c as given by (6.32) and (6.33) were found to be equivalent to 0.0052% and 0.0165%, respectively. Checking on how these figures would reflect on the computed specifications, the results shown in Table 6.11 are obtained.

Example 6.5 Using the semi-analytical trial-and-error approach, redesign the three-phase power transformer of Example 6.2.

Solution

For the steel material under consideration, the specific core loss c_{Fe} fitted in accordance with (4.73) may be assumed to be 0.2644.

Table 6.11 Comparison between required and computed leading specifications using the semi-analytical trial-and-error approach for the single-phase power transformer of Example 6.1

Lead specifications	Required	Achieved after 300 iterations	Error %
Rating S	7.5 MW	7.25 MW	3.33
Maximum copper losses P_{cu}	26 KW	25.52 KW	1.85
Maximum no-load losses P_{NL}	6.9 KW	6.51 KW	5.65
Leakage reactance referred to HV side X	60.87 Ω	59.75 Ω	1.84

Taking the same reasonably assumed design values of Table 6.5, and based upon the required specifications and the reasonably assumed design parameters, G_{3-1}, G_{3-2}, G_{3-3}, and G_{3-4} were computed according to (5.9)–(5.12). From (6.16), the optimum value of H_W is computed. Applying the iterative approach given by (6.34)–(6.36) while taking χ to be 0.95 for a limited number of 300 iterations, J_H and B_c evolve as shown in Figs. 6.5 and 6.6, respectively.

After the trial-and-error iterations, estimates for D, J_H and B_c that would satisfy the design requirements should be found. All other design details may then be computed based upon the detailed analyses and derivations given in Chap. 4. Computed results are given in Table 6.12 for the laminated core material under consideration.

It should be pointed out that after the 300 implemented iterations, the accuracy figures of J_H and B_c as given by (6.32) and (6.33) were found to be equivalent to 0.00022% and 0.00078%, respectively. Checking on how these figures would reflect on the computed specifications, the results shown in Table 6.13 are obtained.

Fig. 6.5 Evolution of the HV side current density (corresponding to the 0.23 mm H-0 material) over the first 300 trial-and-error iterations for the three-phase transformer of Example 6.2

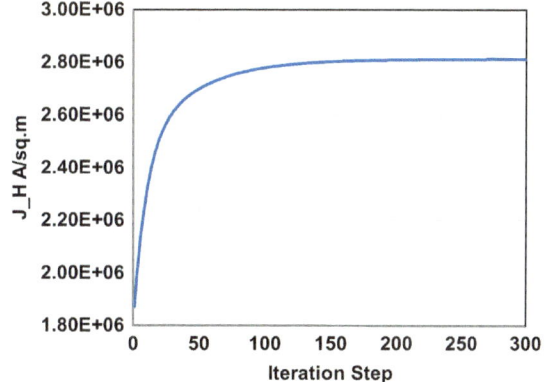

Fig. 6.6 Evolution of the core flux density (corresponding to the 0.23 mm H-0 material) over the first 300 trial-and-error iterations for the three-phase transformer of Example 6.2

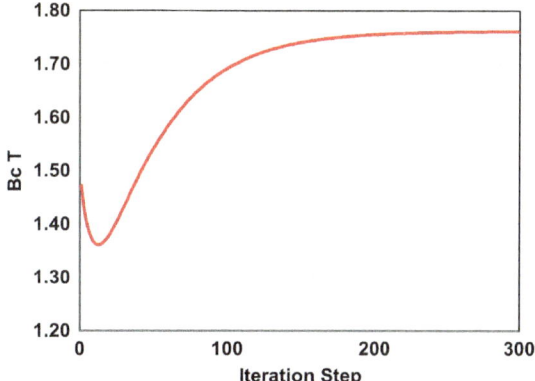

Table 6.12 Computed design details using the semi-analytical trial-and-error approach for the power transformer case of Example 6.2

Variable	Computed value for 0.23 mm H-0
G_{3-1}	10.5554483
G_{3-2}	9.65465E−09
G_{3-3}	1879.512568
G_{3-4}	5.93843E−12
H_w (m)	1.152
D (m)	0.537
J_H (A/m^2)	2,810,505.52
B_c (T)	1.760
J_L (A/m^2)	2,342,087.936
W_w (m)	0.562112187
N_H	819
N_L	79
Wt_{Fe} (Kg)	14,078.69
Wt_{Cu} (Kg)	4308.06

Table 6.13 Comparison between required and computed leading specifications using the semi-analytical trial-and-error approach for the three-phase power transformer of Example 6.2

Lead specifications	Required	Achieved after 300 iterations	Error %
Rating S	20 MW	20,000,201.00 W	0.001
Maximum copper losses P_{cu}	80 KW	79,996.77 W	0.004
Maximum no-load losses P_{NL}	15 KW	14,997.89 W	0.014
Leakage reactance referred to HV side X	65.34 Ω	65.337 Ω	0.005

Example 6.6 Using the semi-analytical trial-and-error approach, redesign the three-phase power transformer of Example 6.3.

Solution

For the steel material under consideration, the specific core loss c_{Fe} fitted in accordance with (4.73) may be assumed to be 0.2644.

Taking the same reasonably assumed design values of Table 6.8, and based upon the required specifications and the reasonably assumed design parameters, G_{3-1}, G_{3-2}, G_{3-3}, and G_{3-4} were computed according to (5.9)–(5.12). From (6.16), the optimum value of

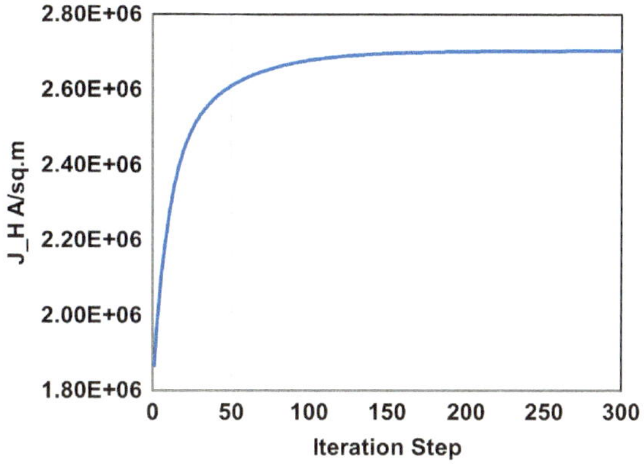

Fig. 6.7 Evolution of the HV side current density (corresponding to the 0.23 mm H-0 material) over the first 300 trial-and-error iterations for the three-phase transformer of Example 6.3

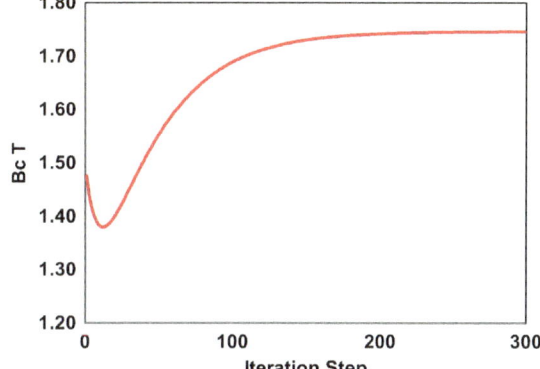

Fig. 6.8 Evolution of the core flux density (corresponding to the 0.23 mm H-0 material) over the first 300 trial-and-error iterations for the three-phase transformer of Example 6.3

H_W is computed. Applying the iterative approach given by (6.34)–(6.36) while taking χ to be 0.95 for a limited number of 300 iterations, J_H and B_c evolve as shown in Figs. 6.7 and 6.8, respectively.

After the trial-and-error iterations, estimates for D, J_H and B_c that would satisfy the design requirements should be found. All other design details may then be computed based upon the detailed analyses and derivations given in Chap. 4. Computed results are given in Table 6.14 for the laminated core material under consideration.

It should be pointed out that after the 300 implemented iterations, the accuracy figures of J_H and B_c as given by (6.32) and (6.33) were found to be equivalent to 0.00015% and 0.00053%, respectively. Checking on how these figures would reflect on the computed specifications, the results shown in Table 6.15 are obtained.

Table 6.14 Computed design details using the semi-analytical trial-and-error approach for the power transformer case of Example 6.3

Variable	Computed value for 0.23 mm H-0
G_{3-1}	8.253306491
G_{3-2}	8.00217E−09
G_{3-3}	1879.512568
G_{3-4}	2.26824E−12
H_w (m)	1.392520707
D (m)	0.575406615
J_H (A/m^2)	2,702,086.321
B_c (T)	1.746074385
J_L (A/m^2)	2,549,138.039
W_w (m)	0.663105099
N_H	719
N_L	69
Wt_{Fe} (Kg)	18,615.97
Wt_{Cu} (Kg)	5094.80

Table 6.15 Comparison between required and computed leading specifications using the semi-analytical trial-and-error approach for the three-phase power transformer of Example 6.3

Lead specifications	Required	Achieved after 300 iterations	Error %
Rating S	25 MW	25 MW	0.0000
Maximum copper losses P_{cu}	99 KW	98.997 KW	0.0030
Maximum no-load losses P_{NL}	19.51 KW	19.508 KW	0.0103
Leakage reactance referred to HV side X	54.415 Ω	54.414 Ω	0.0018

References

1. Adly, A. A. (2017). A specifications-oriented initial design methodology for power transformers. *Energy Systems, 8,* 285–296.
2. Adly, A. A. (2021). Computer-aided transformer design capacity building. *Transformers Magazine, 8*(1), 72–77.
3. Steel, A. K. (2013). TRAN-COR H grain oriented electrical steels, product data bulletin. *Ohio, USA, 1.*

Unconventional Computer-Aided-Design (CAD) Approaches

In the previous chapters, details about the relations between a power transformer operational specification and its design details—including dimensions, materials used and their specifications, window relative dimensions and area occupied by the copper windings—have been discussed. Moreover, conventional design approaches that guarantee fulfilling a set of required specifications have been presented. In this chapter, however, a couple of unconventional computer-aided-design (CAD) approaches are presented. Namely, this chapter discusses how artificial neural networks (ANNs) and the multi-objective particle swarm optimization (MOPSO) approach may be utilized to design a power transformer while minimizing the expertise required to do so. More details on how to utilize both approaches in unconventional design methodologies of power transformers, together with numerical examples, are presented in the following sections.

7.1 Transformer Design Using Feedforward Artificial Neural Networks

Artificial neural networks attempt to imitate the biological neural connections of the human brain, where they typically consist of a large number of elementary processing units called neurons. As shown in Fig. 7.1, the neuron generally applies a nonlinear function f to the weighted sum of the inputs to produce a single output o [1, 2]. Hence, the output of the single neuron is given by:

$$o = f\left(\sum_{j=0}^{J} w_j y_j\right), \tag{7.1}$$

© The Author(s), under exclusive license to Springer Nature Switzerland AG 2025
A. Adly and S. Abd-El-Hafiz, *Unconventional Performance Oriented Power Transformers Design Methodologies*, Synthesis Lectures on Electrical Engineering,
https://doi.org/10.1007/978-3-031-85221-3_7

Fig. 7.1 A single neuron with
J inputs, one output, and
activation function f

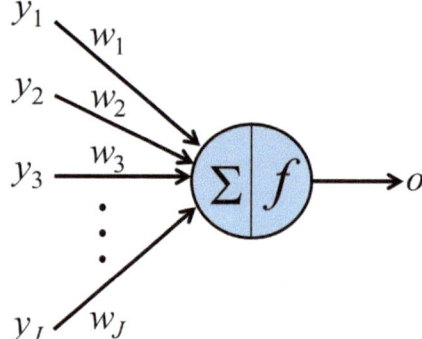

where w_1, w_2, ..., w_J are the weights, y_1, y_2, ..., y_J are the inputs, and f is an activation function. The commonly used activation functions are the step, sigmoid and piecewise linear functions. The sigmoid activation function is given by:

$$f(x) = \frac{1}{1 + e^{-ax}},\tag{7.2}$$

where a is a constant.

In feedforward neural networks (FFNNs), the neurons are arranged in layers and data flows unidirectionally from the input layer through the hidden layers(s) towards the output layer. Figure 7.2 shows the most fundamental architecture of a FFNN, which is a three-layer network that includes a single hidden layer. The input layer consists of $(I - 1)$ externally applied inputs z_1, z_2, ..., z_{I-1} augmented with a constant bias node $z_I = -1$. In order to simplify the presentation, the constant bias node is assumed to be included in the input vector and the constant a of the sigmoid activation function is assumed to be unity. Other layers in the FFNN receive their inputs from the nodes in the previous layer as well as from a single constant bias node. Finally, the outputs result from the last layer.

In general, FFNNs are used in many applications that can be broadly classified into function approximation or pattern classification applications. They are trained using a supervised learning procedure called the error backpropagation algorithm [3]. In supervised learning algorithms, the network is trained using data for which the inputs and the corresponding target outputs are given. Given the 3-layer FFNN of Fig. 7.2 with I nodes in the input layer, J neurons in the hidden layer and K neurons in the output layer, the learning procedure utilizes a training set of P points. Each training point is in the form of an input–output pair $(\mathbf{Z}, \mathbf{D})$, where $\mathbf{Z} = [z_1 z_2 \ldots z_I]^t$ is the input vector and $\mathbf{D} = [d_1 d_2 \ldots d_K]^t$ is the corresponding target output vector. For each point in the training set, the learning step proceeds in two phases; the feedforward phase and the backpropagation phase.

Given an input $\mathbf{Z} = [z_1 z_2 \ldots z_I]^t$, the learning step for this training point starts by computing the responses $\mathbf{Y} = \left[y_1 y_2 \ldots y_J\right]^t$ and $\mathbf{O} = [o_1 o_2 \ldots o_K]^t$ of the hidden and

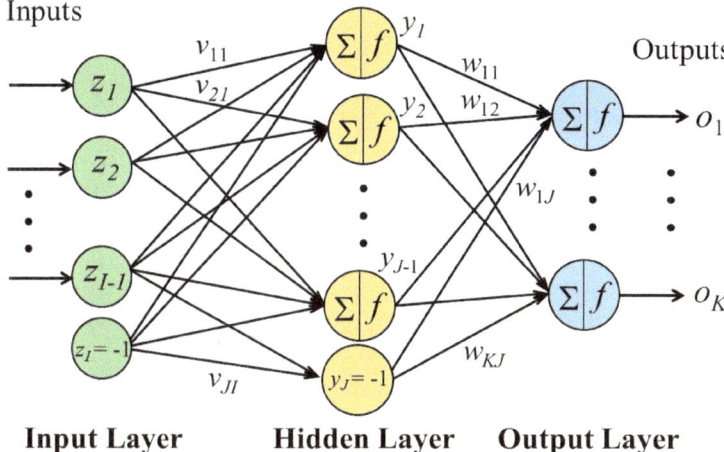

Fig. 7.2   A three-layer feedforward neural network

output layers, respectively, using the following equations of the feedforward phase:

$$Y = F[VZ],\tag{7.3}$$

$$O = F[WY],\tag{7.4}$$

where W is a weight matrix of dimension $K \times J$ with elements w_{kj} connecting the output of neuron j in the hidden layer to the input of neuron k in the output layer. On the other hand, V is a weight matrix of dimension $J \times I$ with elements v_{ji} connecting input node i to the input of neuron j in the hidden layer. $F[.]$ is a nonlinear activation operator represented by a diagonal matrix whose diagonal elements are $f(.)$. The weight matrices are initialized with small random values to decrease the probability of early local convergence.

The objective of the backpropagation phase is to adjust the weight matrices so that the error E in the current training step is minimized:

$$E = 0.5 \sum_{k=1}^{K} (d_k - o_k)^2 = 0.5 \|D - O\|^2,\tag{7.5}$$

where $\|.\|$ is the Euclidean norm. The adjustments of weight matrices W and V are performed using the gradient descent technique in which the individual weight adjustments are given by the following formulas, respectively:

$$\Delta w_{kj} = -\eta \frac{\partial E}{\partial w_{kj}},\tag{7.6}$$

$$\Delta v_{ji} = -\eta \frac{\partial E}{\partial v_{ji}}, \tag{7.7}$$

where η is the learning rate, $i = 1, 2, \ldots, I$, $j = 1, 2, \ldots, J$, and $k = 1, 2, \ldots, K$.

To perform the computations for (7.6) and (7.7), error signal vectors $\boldsymbol{\delta_o}$ and $\boldsymbol{\delta_y}$ are calculated for the output layer and hidden layer neurons, respectively [3]. Elements of the error vector $\boldsymbol{\delta_o}$ are first determined in the output layer, then they are propagated backward towards the hidden layer to calculate the elements of the error vector $\boldsymbol{\delta_y}$:

$$\delta_{ok} = (d_k - o_k)(1 - o_k)o_k, k = 1, 2, \ldots, K, \tag{7.8}$$

$$\delta_{yj} = y_j(1 - y_j) \sum_{k=1}^{K} \delta_{ok} w_{kj}, j = 1, 2, \ldots, J. \tag{7.9}$$

Each element of the vector $\boldsymbol{\delta_o}$ represents the difference between the target and actual output values times the derivative of the activation function. Elements of the vector $\boldsymbol{\delta_y}$ represent products of the activation function derivative times the weighted sum of contributing error signals $\boldsymbol{\delta_o}$ produced by the following layer.

Consequently, the wight matrices W and V are updated to $(W + \Delta W)$ and $(V + \Delta V)$, respectively, using the following two equations:

$$\Delta W = \eta \boldsymbol{\delta_o} Y^t, \tag{7.10}$$

$$\Delta V = \eta \boldsymbol{\delta_y} Z^t. \tag{7.11}$$

In summary, a single training step uses one training point and it consists of the forward phase given by (7.3) and (7.4) and the backpropagation phase described by (7.8) to (7.11). The training step is performed for each point in the training set, and the normalized root mean square error E_{rms} is calculated for all training points $p = 1, 2, \ldots, P$:

$$E_{rms} = \frac{1}{PK} \sqrt{\sum_{p=1}^{P} \sum_{k=1}^{K} (d_{pk} - o_{pk})^2}. \tag{7.12}$$

If E_{rms} for the current training cycle is less than the required threshold E_{max}, training ends successfully. Otherwise, additional training cycles should be performed to reach the required accuracy. Once trained, the final weight matrices are fixed and can be used in testing the obtained network in a feedforward manner. For detailed description and discussion of the error backpropagation algorithm, please refer to [3].

Furthermore, convergence of the error backpropagation algorithm can be accelerated through the use of a momentum update method for the weight matrices [3]. Additional momentum terms are added to (7.10) and (7.11), respectively, as fractions of the most recent weight adjustments:

$$\Delta W(\tau) = -\eta \nabla E(\tau) + \alpha \Delta W(\tau - 1), \tag{7.13}$$

$$\Delta V(\tau) = -\eta \nabla E(\tau) + \alpha \Delta V(\tau - 1), \tag{7.14}$$

where the arguments τ and $(\tau - 1)$ refer to the current and the most recent training step, respectively, and α is a positive momentum constant.

FFNNs are widely used in many engineering applications such as the modeling of complex magnetic media, transformer design and core rewinding (refer, for instance, to [4, 5]). To quantitatively demonstrate how FFNNs may be utilized in power transformers design, please refer to Example 7.1.

Example 7.1 Consider the three-phase ONAN/ONAF 66 kV/11 kV power transformers data given in Table 7.1. Use this data to train an FFNN on correlating the given required specifications to leading design variables, assuming that all transformers under consideration share the design parameters given in Table 7.2. Hence, assess the possibility of utilizing the trained FFNN in designing the two power transformers whose required specifications are:

$S = 20$ MVA, $P_{cu} = 94.5$ KW, $P_{NL} = 13.33$ KW, $X_H = 52.272$ Ω.

$S = 15$ MVA, $P_{cu} = 71$ KW, $P_{NL} = 11$ KW, $X_H = 87.12$ Ω.

Solution

Given the limited number of training data points given in Table 7.1 (i.e., 8 cases), an FFNN having 4 hidden layer nodes was utilized in the training process. Normalized values of the inputs and outputs were used in the training process. More specifically, S, P_{cu}, P_{NL},

Table 7.1 The set of three-phase power transformers data to be utilized in the FFNN training process of Example 7.1

Design #	Operational specifications (inputs)				Design variables (target outputs)			
	S (MW)	P_{cu} (KW)	P_{NL} (KW)	X_H (Ω)	H_W (m)	D (m)	J_H (MA/ m^2)	B_c (T)
1	20	94.5	13.33	65.34	1.159	0.525	2.97	1.722
2	18.75	82	13.20	69.70	1.205	0.515	2.66	1.729
3	18.75	82	13.20	55.76	1.077	0.588	2.89	1.529
4	15	71	11.00	69.70	1.035	0.539	2.89	1.568
5	10	65	7.20	130.68	0.988	0.413	3.21	1.766
6	10	64.8	5.66	104.54	0.884	0.470	3.49	1.569
7	8	60	5.45	163.35	0.920	0.407	3.37	1.600
8	8	60	5.45	130.68	0.823	0.467	3.65	1.406

Table 7.2 Assumed design parameters for all transformers considered in Example 7.1

Window space factor S_w	0.2
Core lamination stacking factor K_c	0.958
Window height to width c_{hw}	2.25
Current density ratio of the HV side to the LV side c_j	1.0
Copper winding resistivity ρ_{cu}	0.000000021
Steel core laminations density δ_{Fe}	7650
Steel core specific loss fitting parameter C_{Fe}	0.2644 for 0.23 mm H-0
Copper winding density δ_{cu}	8933

X_H, H_W, D, J_H and B_c were normalized through dividing them by the values 25,000,000, 100,000, 15,000, 200, 1.5, 0.75, 4,000,000, and 2, respectively. Several training cycles were performed while maintaining a learning rate η and a momentum α equivalent to 4 and 0.6, respectively. By the end of the training process, a normalized root mean square error E_{rms} of 0.0012 was achieved.

In order to test the capability of FFNN to suggest design details for the required two power transformers, the network various weights deduced using the training phase were fixed and suggested H_W, D, J_H and B_c values were generated using the trained FFNN. Operational specifications were then computed using the analytical methodology presented in Chap. 6. Suggested values for the aforementioned variables as well as a comparison between required and achieved performance specifications are given in Table 7.3. Please note that for the example under consideration, the achieved average and maximum percentage errors were found to be equivalent to 5.9% and 18.4%, respectively. These error figures are acceptable due to the limited number of training data points available. Obviously, as the number of training data points increase, it is possible to achieve lower error figures.

The results given in Table 7.3 clearly suggest that, in the absence of any expertise related to power transformers design, FFNNs may be utilized to provide design detail suggestions to meet certain performance specifications. This approach, however, is dependent on the availability of some data correlating performance to design details for a number of actual power transformers. Selection of the number and domain of the training data as well as its validity for the required design ratings would definitely affect the accuracy of this approach.

Table 7.3 Suggested values for the design variables as well as a comparison between required and achieved performance specifications of Example 7.1

Test case #	FFNN suggested design variables		Operational specifications			Error %
			Spec	Required	Computed	
1	H_W	1.098 m	S	20 MW	20.54 MW	2.70
	D	0.576 m	P_{cu}	94.5 KW	99.66 KW	5.46
	J_H	3.136 MA/m^2	P_{NL}	13.33 KW	12.98 KW	2.63
	B_c	1.546 T	X_H	52.272 Ω	61.88 Ω	18.38
2	H_W	1.140 m	S	15 MW	14.32 MW	4.53
	D	0.466 m	P_{cu}	71 KW	67.28 KW	5.24
	J_H	2.650 MA/m^2	P_{NL}	11 KW	10.94 KW	0.55
	B_c	1.800 T	X_H	87.12 Ω	80.24 Ω	7.90

7.2 Transformer Design Using Multi-Objective Particle Swarm Optimization

Swarm intelligence is a branch of artificial intelligence, which is inspired by the behavior of social swarms in nature. Particle swarm optimization (PSO) is a popular swarm intelligence approach that has been extensively studied, updated and used since its proposal in the mid-nineties by Kennedy and Eberhart [6]. Their original stochastic optimization algorithm is inspired by the social behavior concepts of bird flocks, fish schools or insect swarms. Since its inception, many variants of PSO have been designed and utilized in different applications including wireless networks, power systems, classification, and many more diverse real-world optimization tasks [7, 8].

The single-objective PSO is a population-based heuristic approach, which starts by populating the d-dimensional search space $\mathcal{S} \subset \mathbb{R}^d$ with a set of N_s random solutions called particles. Particle number i in the swarm, $i = 1, 2, \ldots, N_s$, is associated with a set of attributes that include its current position $\boldsymbol{x_i} = [x_{i1} x_{i2} \ldots x_{id}]^t$ and velocity $\boldsymbol{v_i} = [v_{i1} v_{i2} \ldots v_{id}]^t$. The individual experience of the particle is also recorded as the position of the particle's personal best performance $\boldsymbol{P_i} = [P_{i1} P_{i2} \ldots P_{id}]^t$. Experience of the whole swarm is stored as the position of the global best performance among all particles $\boldsymbol{G} = [G_1 G_2 \ldots G_d]^t$. The fitness value of any particle is obtained by evaluating the objective function, f_1, at its current position.

During each optimization cycle, the particles move towards the optimum solution by updating their velocity and position according to the following two equations [9], respectively.

$$v_{ij} = wv_{ij} + c_1 r_1 \left(P_{ij} - x_{ij} \right) + c_2 r_2 \left(G_j - x_{ij} \right), \tag{7.15}$$

$$x_{ij} = x_{ij} + v_{ij}, \tag{7.16}$$

where $i = 1, 2, \ldots, N_s$, $j = 1, 2, \ldots, d$, c_1 and c_2 are positive constants, r_1 and r_2 are random numbers generated uniformly in the interval $(0, 1)$, and w is the inertia weight.

The right-hand side of (7.15) consists of three terms representing the momentum, cognitive and social components, respectively. The momentum term moves the particle in the direction it has traveled so far. The cognitive component considers the individual experience of the particle and attracts it towards the best position it has visited so far. The social component represents the interaction between the particles and attracts it towards the best position found by the swarm. Figure 7.3 depicts the motion of a particle from its current position x_i to the new position x_i^{new} in terms of the aforementioned three components. Upon termination of the algorithm, G serves as the solution.

Proper selection of PSO parameters can greatly affect its performance. The random numbers r_1 and r_2 are responsible for providing randomness to the flight of the swarm. The constants c_1 and c_2 are the acceleration coefficients whose values control the contribution of the cognitive and social terms, respectively. Suitable selection of w provides balance between global exploration and local exploitation. There are several strategies for the selection of the inertia weight. Originally, w was decreased linearly from about 0.9 to 0.4 during a run so that the search changes gradually from exploration to exploitation.

Fig. 7.3 The three components of motion in the PSO algorithm

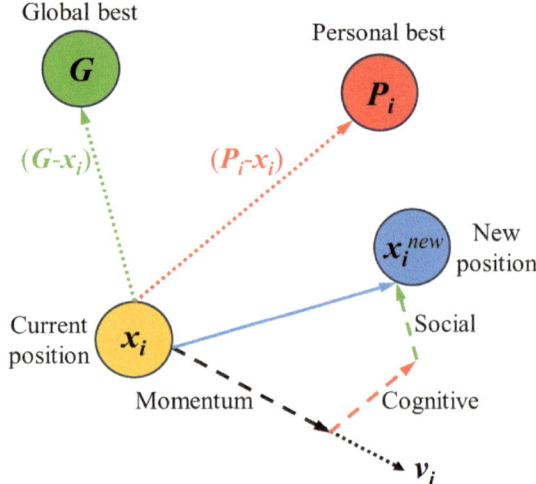

Other approaches included the use of a static w, random w or to adapt w using a fuzzy system [9].

Single-objective PSO has been successfully utilized in many engineering applications such as the optimization of devices and systems [10, 11], field computation in nonlinear magnetic media [12, 13], and robotics [14]. However, engineering design problems usually build computational models that describe complex behaviors of physical systems, and optimal solutions are required with respect to multiple performance criteria. Therefore, multi-objective optimization becomes useful in such scenarios where the designer can select the suitable alternative from a set of optimal compromise solutions.

For multi-objective optimization, there are k objective functions defined over the d-dimensional search space $\mathcal{S} \subset \mathbb{R}^d$:

$$f(x) = [f_1(x)f_2(x) \ldots f_k(x)]^t, \tag{7.17}$$

subject to m inequality constraints:

$$g_i(x) \leq 0, i = 1, 2, \ldots, m. \tag{7.18}$$

It is required to find a solution $x^* = [x_1^* x_2^* \ldots x_d^*]^t$ that minimizes $f(x)$. Optimality of the solution is defined differently because the objective functions $f_i(x)$ may be conflicting with each other. It may be impossible to find a single global minimum at the same point in $\mathcal{S}$. Hence, the notions of Pareto dominance, Pareto optimal set $\mathcal{P}^*$ and Pareto front $\mathcal{PF}^*$ are defined [15].

The Pareto dominance property is defined between two vectors $u = [u_1 u_2 \ldots u_k]^t$ and $v = [v_1 v_2 \ldots v_k]^t$ in the objective space. The vector v is said to dominate the vector u if and only if $v_i \leq u_i$ for all $i = 1, 2, \ldots, k$ and $v_i < u_i$ for at least one component, where k is the dimension of the objective space.

A solution x in the search space $\mathcal{S}$ is said to be Pareto optimal, or nondominated, if and only if there exists no other solution y in $\mathcal{S}$ such that $f(y)$ dominates $f(x)$. The set of all such nondominated solutions constitutes the Pareto optimal set $\mathcal{P}^*$. The Pareto front is defined as the set:

$$\mathcal{PF}^* = \{f(x) : x \in \mathcal{P}^*\}. \tag{7.19}$$

Considerable research has been performed on multi-objective PSO where the single-objective PSO is adapted using miscellaneous techniques such as exploiting each objective function separately or utilizing the Pareto concepts (refer, for instance, to [16–18]). The Pareto dominance concept determines the best positions that guide the swarm during the search procedure. In addition, some criteria are used to improve issues like swarm diversity and Pareto front spread. The time variant multi-objective PSO (TV-MOPSO) algorithm is a Pareto-based algorithm that is adaptive in nature, with respect to the inertia weight and acceleration coefficients, to achieve good balance between exploration and exploitation of the search space [19]. A mutation operator is incorporated to solve the problem of premature convergence, which occurs frequently in multi-objective PSO. An archive is also maintained to store the nondominated solutions found during the search, and the global best solution is selected from this archive using a diversity parameter.

The procedure for implementing the TV-MOPSO algorithm is summarized using the following three main steps [19], which are the initialization, main iteration cycle and output.

1. Initialize the swarm and archive as follows:
   - Randomly generate a swarm SW_0 of size N_s
   - Initialize an archive A_0 of maximum size N_a
2. For $\tau = 1$ to maximum number of iterations M, update the swarm SW_τ, update the archive A_τ, and mutate the swarm SW_τ using the following three sub-steps, respectively:
   (a) For $i = 1$ to N_s, update each particle in the swarm SW_τ using:
      - Get the global best solution, G, from the archive A_τ
      - Get the personal best solution P_i
      - Adjust the time variant parameters w, c_1 and c_2
      - Update the velocity v_{ij} using (7.15), $j = 1, 2, \ldots, d$
      - Update the position x_{ij} using (7.16), $j = 1, 2, \ldots, d$
   (b) Update the archive A_τ
   (c) Mutate the swarm SW_τ
3. Return the Pareto optimal set $\mathcal{P}^* = A_\tau$

The initialization step generates an initial swarm SW_0 of particles having zero velocities and random values for the positions. The initial archive A_0 includes the nondominated solutions from SW_0. The second step represents the main iteration cycle in which the swarm is updated, the archive is updated, and the swarm is mutated at each iteration τ.

Every iteration τ of the main cycle starts, in step 2.a, by updating each particle in the swarm SW_τ. The global best solution G is selected from the archive based on the diversity of the solutions, where the diversity computation uses the nearest neighbor concept. The personal best solution P_i is also fetched. Moreover, the time variant parameters, which are the inertia weight w and the acceleration coefficients c_1 and c_2, are adjusted. Finally, the velocity and position of each particle in the swarm are updated using (7.15) and (7.16), respectively. If the currently obtained solution dominates the personal best solution P_i, then the personal best solution is replaced.

In the initial stages of the search, more global exploration is needed because there is little knowledge about the search space. In later stages, however, more local exploitation is needed to make use of the gained knowledge. Since higher values of the inertia weight w enable global search while lower values support local search, w is decreased linearly from an initial value w_i to a final value w_f. The value of w at iteration τ is calculated as:

$$w = w_f + \frac{(M - \tau)}{M}(w_i - w_f), \tag{7.20}$$

where M is the maximum number of iterations.

To compromise between exploration and exploitation of the search space, the cognitive acceleration coefficient c_1 and the social acceleration coefficient c_2 are also varied linearly with each iteration. Higher values of c_1 enable larger deviation in the search space, while higher values of c_2 support convergence to the global best. Hence, c_1 is decreased from the initial value c_{1i} to the final value c_{1f} according to (7.21), and c_2 is increased from c_{2i} to c_{2f} according to (7.22).

$$c_1 = c_{1i} + \frac{\tau}{M}(c_{1f} - c_{1i}), \tag{7.21}$$

$$c_2 = c_{2i} + \frac{\tau}{M}(c_{2f} - c_{2i}). \tag{7.22}$$

In step 2.b, the archive A_τ is updated by including the nondominated solutions from the combined population of the swarm and the archive ($SW_\tau \cup A_\tau$). If the size of the archive exceeds the maximum limit N_a, it is truncated using the diversity parameter [19].

In step 2.c, a mutation operator is used to improve exploration of the search space while maintaining better diversity. Given a particle p, a randomly selected coordinate of the particle is mutated from p_k to p'_k using:

$$p'_k = \begin{cases} p_k + \Delta(\tau, p_{ku} - p_k), & \text{if } flip = 0 \\ p_k - \Delta(\tau, p_k - p_{kl}), & \text{if } flip = 1 \end{cases}, \tag{7.23}$$

where $flip$, p_{kl} and p_{ku} denote the random event of returning 0 or 1, the lower limit of p_k and the upper limit of p_k, respectively. The function Δ is defined by:

$$\Delta(\tau, x) = \left(1 - r^{(1 - \frac{\tau}{M})^q}\right)x, \tag{7.24}$$

where r is a random number in the range [0, 1], and the parameter q determines the level of mutation's dependence on the iteration number τ.

After executing the specified number of iterations, step 3 concludes the algorithm with the archive containing the Pareto optimal set $\mathcal{P}^*$. For detailed description and discussion of the TV-MOPSO algorithm, please refer to [19].

A bold example of the validity of utilization of this approach in the design of three-phase power transformers has been presented in [20]. In this paper, the approach was utilized to design a power transformer whose required specifications are as given in Example 7.2.

Table 7.4 Required specifications for the three-phase power transformer of Example 7.2 [20]

Rating S	40 MV
High voltage phase value V_{phH}	66 kV
Low voltage phase value V_{phL}	6.35 kV
Frequency f	50 Hz
Maximum copper losses P_{cu}	135.90 KW
Maximum no-load losses P_{NL}	24.70 KW
Leakage reactance referred to HV side X	35.937 Ω
Core laminations type	AK Steel Tran-Cor Grain Oriented steel 0.23 mm H-0

Example 7.2 Using the MOPSO, it is required to design the actual 40 MVA ONAN/ONAF Star/Delta power transformer whose specifications are given in Table 7.4. Compare between the computed and actual design detail values [20].

Solution

For the steel material under consideration, it has been shown in Fig. 6.1 that the specific core loss may be fitted in accordance to (4.73) by taking the value of c_{Fe} to be equivalent to 0.2644. By matching some of the design specifics for the actual power transformer under consideration, reasonable design parameters were assumed as given in Table 7.5 [20].

Based upon the required specifications and the reasonably assumed design parameters, G_{3-1}, G_{3-2}, G_{3-3}, and G_{3-4} were computed according to (5.9)–(5.12) and H_W was determined from (6.16) as given in Table 7.6.

Table 7.5 Reasonably assumed (or extracted from the actual transformer details) design parameters [20]

Window space factor S_w	0.20
Core lamination stacking factor K_c	0.958
Window height to width c_{hw}	2.05
Current density ratio of the HV side to the LV side c_j	1.00
Copper winding resistivity ρ_{cu}	0.000000021
Steel core laminations density δ_{Fe}	7650
Steel core specific loss fitting parameter C_{Fe}	0.2644 for 0.23 mm H-0
Copper winding density δ_{cu}	8933

Table 7.6 Computed values of G_{3-1}, G_{3-2}, G_{3-3}, and G_{3-4} as well as H_W for the three-phase power transformer case of Example 7.2

Variable	Computed value
G_{3-1}	11.61099313
G_{3-2}	1.15856E$-$08
G_{3-3}	1879.512568
G_{3-4}	1.7964E$-$12
H_W (m)	1.31

Having deduced H_W, and referring to (5.13)–(5.16), the multi-objective optimization technique may then be utilized to determine D, J_H and B_c that would result in meeting the design specifications requirements.

Let the required solution and objective function be given by:

$$x^* = \left[J_H^* \ B_c^* \ D^* \right]^t, \tag{7.25}$$

$$f(x) = \left[\left(\frac{S^c(x^*) - S}{2} \right)^4 \ \left(P_{cu}^c(x^*) - P_{cu} \right)^4 \ \left(\frac{P_{NL}^c(x^*) - P_{NL}}{4} \right)^4 \right]^t, \tag{7.26}$$

where, S^c, P_{cu}^c and P_{NL}^c represent computed values.

Hence, the multi-objective optimization problem reduces to the determination of x^* that leads to a minimum value of the function $f(x)$ while maintaining reasonable values for all three unknown variables D, J_H and B_c (i.e., subject to range constraints).

From a practical design perspective, the unknown variables range constraints may be defined by [20]:

$$\begin{bmatrix} 1.1 \times 10^6 \\ 1.0 \\ 0.1 \end{bmatrix} \leq \begin{bmatrix} J_H \\ B_c \\ D \end{bmatrix} \leq \begin{bmatrix} 3.2 \times 10^6 \\ 1.7 \\ 0.7 \end{bmatrix}. \tag{7.27}$$

The previously described TV-MOPSO algorithm was implemented for the power transformer design case under consideration while using a swarm of particles M_s, a maximum archive size M_a, and maximum number of iterations M of 50, 200, and 1000, respectively. Within those iterations, the implementation parameters w_i, w_f, c_{1i}, c_{1f}, c_{2i}, c_{2f}, and q were set to 0.7, 0.4, 2.5, 0.5, 0.5, 2.5, and 5, respectively. The Pareto front obtained for this case is shown in Fig. 7.4, and a comparison between computed and actual transformer lead design and performance figures are given in Table 7.7.

It can be seen from the results given in Table 7.7 that the MOPSO technique can be utilized in power transformers design tasks. Design details suggested by this technique can obviously be used as a very good initial guess to be refined by more sophisticated electromagnetic field techniques.

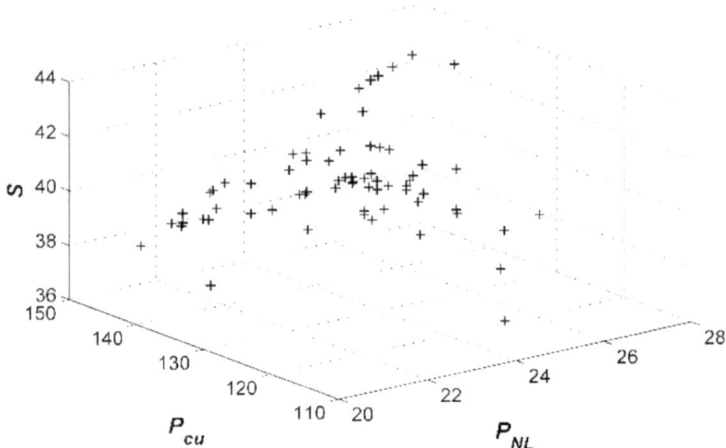

Fig. 7.4 Obtained Pareto front for the 40 MVA three-phase power transformer of Example 7.2 [20]

Table 7.7 Comparison between actual and computed lead design and performance figures for the 40 MVA three-phase power transformer of Example 7.2 [20]

Lead specifications	Actual value	Computed value
Rating S	40 MW	40.24 MW
Maximum copper losses P_{cu}	135.90 KW	135.98 KW
Maximum no-load losses P_{NL}	24.70 KW	24.17 KW
Leakage reactance referred to HV side X	35.937 Ω	35.970 Ω
H_W (m)	1.37	1.37
D (m)	0.61	0.64
J_H (A/m^2)	2,170,000.00	2,580,000.00
B_c (T)	1.75	1.74
Wt_{Fe} (Kg)	20,425.5	22,797.0
Wt_{Cu} (Kg)	9111.7	7235.7

References

1. Aggarwal, C. C. (2023). *Neural networks and deep learning* (2nd ed.). Springer International Publishing.
2. Annema, A.-J. (1995). *Feed-forward neural networks*. Springer, US.
3. Zurada, J. (1992). *Introduction to artificial neural systems*. West Publishing Co.
4. Adly, A. A., & Abd-El-Hafiz, S. K. (1999). Automated transformer design and core rewinding using neural networks. *Journal of Engineering and Applied Science, 46*, 351–364.

5. Adly, A. A., & Abd-El-Hafiz, S. K. (2014). Utilizing neural networks in magnetic media modeling and field computation: A review. *Journal of advanced research, 5*(6), 615–627.

6. Kennedy, J., & Eberhart, R. (1995). Particle swarm optimization. In *Proceedings of ICNN'95-international conference on neural networks* (Vol. 4, pp. 1942–1948). IEEE.

7. Gad, A. G. (2022). Particle swarm optimization algorithm and its applications: A systematic review. *Archives of computational methods in engineering, 29*(5), 2531–2561.

8. Tang, J., Liu, G., & Pan, Q. (2021). A review on representative swarm intelligence algorithms for solving optimization problems: Applications and trends. *IEEE/CAA Journal of Automatica Sinica, 8*(10), 1627–1643.

9. Eberhart, R.C., & Shi, Y. (2001). Particle swarm optimization: developments, applications and resources. In *Proceedings of the 2001 congress on evolutionary computation (IEEE Cat. No. 01TH8546)* (Vol. 1, pp. 81–86). IEEE.

10. Adly, A. A., & Abd-El-Hafiz, S. K. (2006). Using the particle swarm evolutionary approach in shape optimization and field analysis of devices involving nonlinear magnetic media. *IEEE Transactions on Magnetics, 42*(10), 3150–3152.

11. Adly, A. A., & Abd-El-Hafiz, S. K. (2007). Speed-range-based optimization of nonlinear electromagnetic braking systems. *IEEE transactions on magnetics, 43*(6), 2606–2608.

12. Adly, A. A., & Abd-El-Hafiz, S. K. (2009). Utilizing particle swarm optimization in the field computation of nonlinear media subject to mechanical stress. *Journal of Applied Physics, 105*(7), 07D507.

13. Adly, A. A., & Abd-El-Hafiz, S. K. (2004). Field computation in non-linear magnetic media using particle swarm optimization. *Journal of Magnetism and Magnetic Materials, 272*, 690–692.

14. Adly, M. A., & Abd-El-Hafiz, S. K. (2016). Inverse kinematics using single-and multi-objective particle swarm optimization. In *2016 28th international conference on microelectronics (ICM)* (pp. 269–272). IEEE.

15. Parsopoulos, K. E., & Vrahatis, M. N. (2008). Multi-objective particles swarm optimization approaches. In *Multi-objective optimization in computational intelligence: Theory and practice* (pp. 20–42). IGI global.

16. Trivedi, V., Varshney, P., & Ramteke, M. (2020). A simplified multi-objective particle swarm optimization algorithm. *Swarm Intelligence, 14*(2), 83–116.

17. Cui, Y., Meng, X., & Qiao, J. (2022). A multi-objective particle swarm optimization algorithm based on two-archive mechanism. *Applied soft computing, 119*, 108532.

18. Han, H., Liu, Y., Hou, Y., & Qiao, J. (2023). Multi-modal multi-objective particle swarm optimization with self-adjusting strategy. *Information Sciences, 629*, 580–598.

19. Tripathi, P. K., Bandyopadhyay, S., & Pal, S. K. (2007). Multi-objective particle swarm optimization with time variant inertia and acceleration coefficients. *Information sciences, 177*(22), 5033–5049.

20. Adly, A. A., & Abd-El-Hafiz, S. K. (2015). A performance-oriented power transformer design methodology using multi-objective evolutionary optimization. *Journal of Advanced Research, 6*(3), 417–423.

Conclusions

There is no doubt that power transformers are regarded among the indispensable components in any power network. Thus, optimized and efficient transformer designs having preset specifications are crucial for the minimization of the overall network capital and operational costs. As previously stated, most of the relevant literature related to power transformers design methodologies can be broadly classified into two categories. The first category includes methodologies that adopt a mixed approach of approximate analytical and empirical design equations. The second category includes methodologies that are more accurate yet require expensive software tools, such as FEA packages, and/or massive computation resources. In order to make this book self-contained to a great extent, important characteristics of the main common materials used in fabricating power transformers, such as conductors, steel laminations, and insulators, were reported. Moreover, important basics related to power transformers theory of operation and design aspects, such as Ampere's law, Faraday's law, Lenz's law, the magnetic circuit concept, differences between DC and AC magnetic circuits, approximation of the core *B-H* characteristics, the transformer equivalent circuit, and cooling requirements, were reviewed. The main contribution of this book is its detailed presentation of computationally efficient and reasonably accurate specifications-oriented initial design methodologies that incorporate some guidelines inferred from relevant electromagnetic field analysis studies.

Based upon the detailed analyses, discussions and examples presented in the previous chapters of this book, the following points have to be stressed:

- An important aspect of the proposed design methodologies is the possibility of correlating the main specification requirements, given by the transformer volt-ampere rating, the overall load losses, the no-load losses, and the ohmic reactance per phase, to the

© The Author(s), under exclusive license to Springer Nature Switzerland AG 2025
A. Adly and S. Abd-El-Hafiz, *Unconventional Performance Oriented Power Transformers Design Methodologies*, Synthesis Lectures on Electrical Engineering,
https://doi.org/10.1007/978-3-031-85221-3_8

transformer lead design variables given by the window height, the limb diameter, the maximum core flux density, and the high voltage winding current density. It has also been shown that all other design details may be deduced from the aforementioned design variables (see, for instance, [1]).

- While following the design reasoning of the proposed design methodologies, the values of several design parameters involved in the process, such as resistivity, window height to width ratio, ... etc., have to be set. Those variables may either be identified from the utilized materials characteristics or within ranges inferred from previous analytical and/or electromagnetic field computations (refer, for instance, to [2] and [3]).

- Since the proposed methodology involves worst case scenarios suggesting that the overall load losses and the no-load losses are assumed to be 120% of the ohmic losses and 130% of the iron losses, respectively, the achieved design might yield less losses than the maximum values of the required design. This usually leads to an excess in the materials usage and highlights the role of using more sophisticated FEA tools to accurately assess the overall losses.

- Procedures to implement the proposed design methodologies, using conventional computer aided analytical and semi-analytical trial-and-error approaches, have been presented in detail. It can be deduced from the included numerical examples that, despite the computationally efficient nature of the proposed methodologies, reasonably accurate power transformer designs meeting a set of required specifications may be achieved.

- Procedures to implement the proposed design methodologies, using unconventional computer-aided MOPSO, have been presented in detail. In addition, details of an unconventional computer-aided design approach using FFNN have also been presented. Numerical examples related to these unconventional approaches clearly highlight their validity to meet a set of design specifications with reasonable accuracy (see, for instance, [4] and [5]). It should be pointed out that utilization of the proposed unconventional design approaches could minimize the required expertise for designing power transformers.

Overall, this book has presented useful, practical and computationally efficient approaches, which may be utilized in designing power transformers that meet a required set of specifications. Given the computational efficiency of these approaches, they may be easily incorporated in other optimization tools to investigate material characteristics versus cost tradeoff scenarios. In all cases, the design details achieved by the proposed methodologies of this book may be regarded as very good initial guesses to be refined, if necessary, by more sophisticated computational tools.

References

1. Adly, A. A. (2017). A specifications-oriented initial design methodology for power transformers. *Energy Systems*, *8*, 285–296.2.
2. Saleh, A., Adly, A., Fawzi, T., Omar, A., & El-Debeiky, S. (2002). Estimation and minimization techniques of eddy current losses in transformer windings. In *Proceedings of the CIGRE conference, Paris, France* (pp. 1–6).
3. Saleh, A., Omar, A., Amin, A., Adly, A., Fawzi, T., & El-Debeiky, S. (2004). Estimation and minimization techniques of transformer tank losses. In *Proceedings of the CIGRE conference, Paris, France* (pp. 1–6).
4. Adly, A. A., & Abd-El-Hafiz, S. K. (1999). Automated transformer design and core rewinding using neural networks. *Journal of Engineering and Applied Science, 46*, 351–364.
5. Adly, A. A., & Abd-El-Hafiz, S. K. (2014). Utilizing neural networks in magnetic media modeling and field computation: A review. *Journal of advanced research, 5*(6), 615–627.